Holt Mathematics

Chapter 3 Resource Book

HOLT, RINEHART AND WINSTON
A Harcourt Education Company
Orlando • Austin • New York • San Diego • London

Printed in the United States of America

ISBN 0-03-078193-0

7 8 9 10 170 10 09

CONTENTS

Blackline Masters

Date ___________

Dear Family,

In this chapter, your child will learn to read, write, and order decimals and to solve problems involving decimals. Your child will also learn how to write large whole numbers in scientific notation, which is a shorthand method for writing large numbers.

It is essential that your child be able to work with decimals, since computers and calculators use decimal representation. Since scientific notation is used in the sciences, it is important that your child has a good understanding of how to work with it.

A decimal number can be expressed in three different forms.

Words: sixteen and nine hundredths
Standard form: 16.09
Expanded form: 10 + 6 + 0.0 + 0.09

Students learn to **round** decimals in order to **estimate** sums and differences.

For example, in the problem 3.92 + 6.48:

1. Round 3.92 to the nearest whole number. ⟶ 4

2. Round 6.48 to the nearest whole number. ⟶ 6

3. 4 + 6 = 10 The sum is about 10.

The following are examples of mathematical operations using decimals.

Addition: 9.800 + 9.700 + 9.425 + 9.850

$$\begin{array}{r} 9.800 \\ 9.700 \\ 9.425 \\ +\ 9.850 \\ \hline 38.775 \end{array}$$

In order to solve:
Place the numbers in a column, aligning the decimal points. Use zeros for placeholders.
Add and place the decimal point.

Subtraction: 10 − 9.85

$$\begin{array}{r} {\scriptstyle 9\,9\,10} \\ 1\not{0}.\not{0}\not{0} \\ -9.85 \\ \hline 0.15 \end{array}$$

In order to solve:
Place the numbers in a column, aligning the decimal points. Use zeros as placeholders.
Subtract and place the decimal point.

Multiplication: 0.17 × 3

$$\begin{array}{r} 0.17 \\ \times\ 3 \\ \hline 0.51 \end{array}$$

In order to solve:
Multiply the numbers as you would in whole number multiplication. In order to place the decimal point, count the number of decimal places in each number of the problem. There are two decimal places in 0.17 and none in 3. That is a total of 2 places, so the decimal point will be moved two places to the left.

Division of a decimal number by a whole number: 0.75 ÷ 5

$$\begin{array}{r} .15 \\ 5\overline{)0.75} \\ -5 \\ \hline 25 \\ -25 \\ \hline 0 \end{array}$$

In order to solve:
Place a decimal point in the quotient directly above the decimal point in the dividend.
Divide as you would with a whole number.

Division of a decimal number by a decimal: 3.6 ÷ 1.2

$$1.2\overline{)3.6}$$

$$\begin{array}{r} 3. \\ 1.2.\overline{)3.6.} \\ -36. \\ \hline 0 \end{array}$$

In order to solve:
Make the divisor a whole number by multiplying the divisor and dividend by the same power of 10. 1.2 multiplied by 10 becomes 12 and 3.6 multiplied by 10 becomes 36. Then divide as you would with a whole number.

A number written in scientific notation has two numbers that are multiplied.

$$5.174 \times 10^4$$

The first part is a number that is greater than 1 and less than 10.

The second part is a power of 10.

To write 5.174×10^4 in standard form:

5.174×10^4 The power of 10 is 4.

5.1740. Move the decimal point 4 places right.
Use zeros as placeholders.

$5.174 \times 10^4 = 51,740$

For additional resources, visit go.hrw.com and enter the keyword MR7 Parent.

Name ______________________ Date __________ Class __________

LESSON 3-1

Practice A

Representing, Comparing, and Ordering Decimals

Write the value of the underlined digit in each number.

1. $1.\underline{6}$ ______________

2. $\underline{7}.62$ ______________

3. $3.6\underline{9}$ ______________

4. $\underline{2}0.4$ ______________

5. $5.\underline{1}36.$ ______________

6. $5.0\underline{8}$ ______________

Write each decimal in standard form, expanded form, and words.

7. 1.8 ______________________________

8. 3 + 0.6 + 0.02 ______________________________

9. one and fifty-two hundredths ______________________________

Circle the letter of the correct answer.

10. Which of the following sets is written in order from greatest to least?

A 1.7, 1.07, 17

B 5.2, 2.5, 0.52

C 1.07, 17, 1.7

D 2.5, 0.52, 5.2

11. Which of the following sets is written in order from least to greatest?

F 0.85, 8.5, 5.8

G 4.3, 3.4, 0.43

H 5.8, 0.85, 8.5

J 0.43, 3.4, 4.3

12. Reno, Nevada, gets an average of only five-tenths inch of rain in June, and only three-tenths inch of rain in July. Which month in Reno has less rain?

____

13. Honolulu, Hawaii, gets an average of three and eight tenths inches of rain in December, and three and six tenths inches of rain in January. Which month in Honolulu has more rain?

____

Name ______________________ Date ____________ Class ____________

LESSON 3-1

Practice B
Representing, Comparing, and Ordering Decimals

Write each decimal in standard form, expanded form, and words.

1. 2.07 ____________________

2. 5 + 0.007 ____________________

3. four and six tenths ____________________

4. sixteen and five tenths ____________________

5. 9 + 0.6 + 0.08 ____________________

6. 1.037 ____________________

7. 2 + 0.1 + 0.003 ____________________

8. eighteen hundredths ____________________

9. 6.11 ____________________

Order the decimals from least to greatest.

10. 3.578, 3.758, 3.875

____

11. 0.0943, 0.9403, 0.9043

____

12. 12.97, 12.957, 12.75

____

13. 1.09, 1.901, 1.9, 1.19

____

14. Your seventh and eighth ribs are two of the longest bones in your body. The average seventh rib is nine and forty-five hundredths inches long, and the average eighth rib is 9.06 inches long. Which bone is longer?

____

15. The average female human heart weighs nine and three tenths ounces, while the average male heart weighs eleven and one tenth ounces. Which human heart weighs less, the male or the female?

____

16. The state has $42.3 million for a new theater. The theater that an architect designed would cost $42.25 million. Can the theater be built for the amount the state can pay?

____

17. Lyn traveled 79.47 miles on Saturday, 54.28 miles on Sunday, 65.5 miles on Monday, and 98.43 miles on Tuesday. Which day did she travel the greatest number of miles?

____

Name ______________________ Date __________ Class __________

LESSON 3-1

Practice C

Representing, Comparing, and Ordering Decimals

Compare. Write <, >, or =.

1. 10.569 ☐ 10 + 0.05 + 0.006 + 0.0009

2. seventy-five hundredths ☐ 7.50

3. twenty thousandths ☐ twelve hundredths

4. 98.30675 ☐ 90 + 8 + 0.03 + 0.007

Order from least to greatest.

5. 12.8962, 12.9682, 12.8692 ______________________

6. 8.098, 7.098, 8.079, 7.089 ______________________

7. 65.21, 6.521, 6.0521, 65.12 ______________________

8. 0.304, 0.30, 0.403, 0.43, 0.34 ______________________

Order from greatest to least.

9. 9.653, 90.563, 90.6053 ______________________

10. 11.717, 11.771, 11.117, 11.171 ______________________

11. 8.0359, 8.3509, 8.359, 8.5 ______________________

12. 2.35, 2.05, 2.03, 2.30, 2.53, 2.5 ______________________

13. From 1984 to 1996 American Carl Lewis won every men's gold medal for the long jump at the Summer Olympic Games. In 1984, he jumped 8.54 meters. In 1988, he jumped 8.72 meters. In 1992, he jumped 8.67 meters, and in 1996, he jumped 8.5 meters. Write Lewis' long jumps in order from least to greatest distance. In which year did he jump the farthest?

____

14. The world records for the four highest pole vaults are 6.01 meters, 6.03 meters, 6.0 meters, and 6.14 meters. Write the records in order from greatest to least height. What is the world record for the highest pole vault?

____

Name ______________________ Date __________ Class __________

LESSON 3-1

Reteach

Representing, Comparing, and Ordering Decimals

You can use place value to write decimals in standard form, expanded form, and word form.

To write 2.14 in expanded form, write the decimal as an addition expression using the place value of each digit.

Ones	Tenths	Hundredths	Thousandths	Ten Thousandths
2.	1	4		

2.14 can be written as 2 + 0.1 + 0.04.

When you write a decimal in word form, the number before the decimal point tells you how many wholes there are. The decimal point stands for the word "and."

Notice that the place value names to the right of the decimal begin with tenths, hundredths, and then thousandths. The "ths" ending indicates a decimal.

2.14 can also be written as *two and fourteen hundredths.*

1. How would you read a number with 4 decimal places?

____

Write each decimal in standard form, expanded form, and word form.

2.

Ones	Tenths	Hundredths	Thousandths	Ten Thousandths
5.	6	9	8	

____

____

____

3.

Ones	Tenths	Hundredths	Thousandths	Ten Thousandths
0.	0	9	4	

____

____

____

4. 7 + 0.8

____

____

5. twelve-hundredths

____

____

Name ______________________ Date ____________ Class ____________

LESSON **3-1**

Reteach

Representing, Comparing, and Ordering Decimals (cont.)

You can use place value to compare decimals.
Use $<$ or $>$ to compare the decimals.

Ones	Tenths	Hundredths	Thousandths	Ten Thousandths
3.	7	6	8	
3.	7	5	4	

$0.06 > 0.05$, so $3.768 > 3.754$.

Compare. Write $>$, $<$, or $=$.

6.

Ones	Tenths	Hundredths	Thousandths	Ten Thousandths
1.	0	3		
1.	3			

1.03 □ 1.3

7.

Ones	Tenths	Hundredths	Thousandths	Ten Thousandths
4.	6	7		
4.	6	7	0	

4.67 □ 4.670

8.

Ones	Tenths	Hundredths	Thousandths	Ten Thousandths
0.	3	6	4	5
0.	3	4	6	5

0.3645 □ 0.3465

9. 8.53 □ 8.053

10. 2.253 □ 2.1345

11. 0.87 □ 0.08703

You can use place value to order decimals.

To order 9.76, 8.59, and 9.24, from least to greatest, first compare the numbers in pairs.

Ones	Tenths	Hundredths	Thousandths	Ten Thousandths
9.	7	6		
8.	5	9		
9.	2	4		

$9.76 > 8.59$, $8.59 < 9.24$, $9.76 > 9.24$.

So the numbers from least to greatest are 8.59, 9.24, 9.76.

Write the decimals in order from least to greatest.

12. 0.54, 0.43, 0.52

____

13. 3.43, 3.34, 3.4

____

14. 8.9, 9.8, 9.5

____

15. 0.83, 0.8, 0.083

____

16. 1.1, 0.01,1.01

____

17. 6.5, 6.0, 0.6

____

Name ______________________ Date __________ Class __________

LESSON 3-1

Challenge

Place Your Values

Complete the tables below to show different numbers that can be written with the same digits. Do not use the same digit more than once for each place value.

1. Use the digits 1, 3, 5, 7, and 9 to write four 5-digit numbers of increasing value.

Hundreds	Tens	Ones	Tenths	Hundredths	Thousandths	Ten-Thousandths

2. Use the digits 0, 2, 4, 6, 7, and 8 to write four 6-digit numbers of decreasing value.

Hundreds	Tens	Ones	Tenths	Hundredths	Thousandths	Ten-Thousandths

3. Use the digits 0, 1, 2, 3, 4, 5, and 6 to write four 7-digit numbers of increasing value.

Hundreds	Tens	Ones	Tenths	Hundredths	Thousandths	Ten-Thousandths

Name ______________________ Date __________ Class __________

LESSON 3-1

Problem Solving

Representing, Comparing, and Ordering Decimals

Use the table to answer the questions.

1. What is the heaviest marine mammal on Earth?

____

2. Which mammal in the table has the shortest length?

____

3. Which mammal in the table is longer than a humpback whale, but shorter than a sperm whale?

____

Largest Marine Mammals

Mammal	Length (ft)	Weight (T)
Blue whale	110.0	127.95
Fin whale	82.0	44.29
Gray whale	46.0	32.18
Humpback whale	49.2	26.08
Right whale	57.4	39.37
Sperm whale	59.0	35.43

Circle the letter of the correct answer.

4. Which mammal measures forty-nine and two tenths feet long?
   A blue whale
   B gray whale
   C sperm whale
   D humpback whale

5. Which mammal weighs thirty-five and forty-three hundredths tons?
   F right whale
   G sperm whale
   H gray whale
   J fin whale

6. Which of the following lists shows mammals in order from the least weight to the greatest weight?
   A sperm whale, right whale, fin whale, gray whale
   B fin whale, sperm whale, gray whale, blue whale
   C fin whale, right whale, sperm whale, gray whale
   D gray whale, sperm whale, right whale, fin whale

7. Which of the following lists shows mammals in order from the greatest length to the least length?
   F sperm whale, right whale, humpback whale, gray whale
   G gray whale, humpback whale, right whale, sperm whale
   H right whale, sperm whale, gray whale, humpback whale
   J humpback whale, gray whale, sperm whale, right whale

Name ______________________ Date ____________ Class ____________

LESSON **3-1**

Reading Strategies

Connect Symbols and Words

You can read and write decimals in three ways. A place value chart can help you read decimals.
When you read or say a decimal, say "and" when you come to the decimal point.

Ones	Tenths	Hundredths
2.	5	
0.	1	7
8.	0	6

Read:
2 **and** 5 tenths
17 hundredths
8 **and** 6 hundredths

Use this chart to help you write decimals in standard form and in expanded form.

Words and Symbols	Standard Form	Expanded Form
2 and 5 tenths	2.5	2 + 0.5
17 hundredths	0.17	0.1 + 0.07
8 and 6 hundredths	8.06	8 + 0.06

Write each number in words and symbols, standard form, or expanded form.

1. Write 2.17 with words and symbols. ______________________

2. Write 2.17 in expanded form. ______________________

3. Write 3 and 6 hundredths in standard form. ______________________

4. Write 3 and 6 hundredths in expanded form. ______________________

5. Write 1.5 with words and symbols. ______________________

6. Write 1.5 in expanded form. ______________________

Name ______________________ Date __________ Class __________

Puzzles, Twisters & Teasers

LESSON **3-1** *Criss-Cross Riddle*

For each number written in standard form in the left column, there is a matching number in the right column. Draw a line connecting the numbered circles on the left to the circles beside their matches on the right. To solve the riddle, find the letters crossed by the lines from each number and write the letters with that corresponding number in the blanks below.

7.251	 (1)		◯ Nine and sixty-five hundredths
		(T)	
		(P)	
2.5	(2)		◯ Twelve and one hundredth
		(Y)	
65.65	(3)	(E)	◯ 7 + 0.2 + 0.05 + 0.001
		(K)	
9.65	(4)	(O)	◯ 3 + 0.2 + 0.01 + 0.002
		(G)	
3.212	(5)	(L)	◯ four and four tenths
		(I)	
12.01	(6)	(N)	◯ 2 + 0.5
		(F)	
4.4	(7)		◯ sixty-five and sixty-five hundredths

Riddle:

What is another name for a parrot that has flown away?

A ___ ___ ___ ___ ___ ___ ___ !

1 2 3 4 5 6 7

Name ______________________ Date ____________ Class ____________

LESSON 3-2

Practice A

Estimating Decimals

Round each decimal to the underlined place value.

1. $1.\underline{7}8$ ____________

2. $0.5\underline{6}9$ ____________

3. $1\underline{2}.62$ ____________

4. $3.\underline{2}15$ ____________

5. $24.6\underline{0}8$ ____________

6. $\underline{3}7.84$ ____________

Estimate. Round to the indicated place value.

7. 3.67 + 1.23; tenths ____________

8. 0.726 + 0.119; hundredths ____________

9. 12.86 − 5.73; tenths ____________

10. 8.643 − 2.795; nearest whole number ____________

Estimate each product or quotient.

11. 17.6 ÷ 6.2 ____________

12. 1.9 • 7.045 ____________

13. 23.8 ÷ 4.3 ____________

14. 9.02 • 4.65 ____________

15. 36.1 ÷ 3.9 ____________

16. 2.8 • 5.35 ____________

17. Tyler measured the growth of a plant for his science project. When he started the project, the plant was 2.8 inches tall. At the end of the project, the plant was 5.2 inches tall. About how many inches did the plant grow during Tyler's project?

____

18. Latoya bought 16.2 yards of cloth to make costumes for the school play. She needs 3.8 yards of the cloth to make each costume. About how many costumes can Latoya make with the cloth she bought?

____

Name ______________________ Date ____________ Class ____________

LESSON 3-2

Practice B

Estimating Decimals

Estimate by rounding to the indicated place value.

1. 7.462 + 1.809; tenths

____

2. 15.3614 − 2.0573; hundredths

____

3. 56.4059 − 4.837; ones

____

4. 0.60871 + 1.2103; hundredths

____

Estimate each product or quotient.

5. 42.1 ÷ 5.97

____

6. 11.8 • 6.125

____

7. 63.78 ÷ 8.204

____

8. 7.539 • 3.0642

____

9. 80.794 ÷ 8.61

____

10. 19.801 • 2.78

____

Estimate a range for each sum.

11. 6.8 + 4.3 + 5.6

____

12. 12.63 + 9.86 + 20.30

____

13. Two sixth-grade classes are collecting money to buy a present for one of their teachers. One class collected $24.68 and the other class collected $30.25. About how much money did they collect in all? The gift they want to buy costs $69.75. About how much more money do they need?

____

14. On the highway, Anita drove an average speed of 60.2 miles per hour. At that speed, about how far can she travel in three and a half hours? At that same speed, about how many hours will it take Anita to drive 400 miles?

____

Name ______________________ Date __________ Class __________

LESSON 3-2

Practice C
Estimating Decimals

Estimate by rounding to the indicated place value.

1. 102.68 + 57.209; ones

2. 83.56912 − 11.7415; hundredths

3. 215.673 − 9.5511; tenths

4. 99.97452 + 5.6409; tenths

Estimate each product or quotient.

5. 24.674 • 5.578

6. 107.95 ÷ 9.193

7. 72.603 ÷ 10.164

8. 6.694 • 38.69

9. 140.9 ÷ 11.756

10. 48.6035 • 7.63

Estimate a range for each sum.

11. 8.4 + 3.1 + 9.54 + 2.86

12. 17.38 + 5.93 + 18.60 + 29.34

13. For the first half of the year, the average monthly precipitation levels in Atlanta, Georgia, are 4.8 inches, 4.8 inches, 5.8 inches, 4.3 inches, 4.3 inches, and 3.6 inches. Estimate the average precipitation level in Atlanta for January through June. Then find the actual average precipitation level for those months. How does your estimate compare to the actual data?

14. Tammy earns $14.68 per hour working as a chef. On average she works 40.3 hours a week. About how much does Tammy earn each week? About how much does she earn each year?

Name ______________________ Date ____________ Class ____________

LESSON 3-2

Reteach

Estimating Decimals

You can use rounding to estimate. Round to the indicated place value. Then add or subtract.

A. 3.478 + 7.136; tenths

3.<u>4</u>78   7 ≥ 5, so round up   3.5
7.<u>1</u>36   3 < 5, so round down   +7.1
10.6

3.478 + 7.136 is about 10.6.

B. 12.848 − 6.124; hundredths

12.8<u>4</u>**8**   8 ≥ 5, so round up   12.85
6.1<u>2</u>**4**   4 < 5, so round down   −6.12
6.73

12.848 − 6.124 is about 6.73.

Estimate by rounding to the indicated place value.

1. 1.04 + 9.37; tenths

1.04 rounds to ______

9.37 rounds to ______

estimate ______

2. 2.17 + 3.56; tenths

2.17 rounds to ______

3.56 rounds to ______

estimate ______

3. 6.753 − 4.245; hundredths

6.753 rounds to ______

4.255 rounds to ______

estimate ______

You can use compatible numbers to estimate. Pick numbers that are close to the actual numbers that are easy to multiply or divide. Then multiply or divide.

A. 4.6 • 3.2
5 and 3 are compatible numbers.
5 • 3 = 15, so 4.6 • 3.2 is about 15.

B. 48.3 ÷ 13.2
48 and 12 are compatible numbers.
48 ÷ 12 = 4, so 48.3 ÷ 13.2 is about 4.

Use compatible numbers to estimate each product or quotient.

4. 9.4 • 5.6 ______

5. 7.25 • 10.84 ______

6. 84.8 ÷ 3.9 ______

7. 21.9 ÷ 3.1 ______

8. 8.3 • 7.6 ______

9. 55.7 ÷ 6.9 ______

10. 5.57 ÷ 2.7 ______

11. 6.729 • 9.8 ______

Name ______________________ Date ____________ Class ____________

LESSON 3-2

Challenge

Out to Lunch

Use the restaurant bills below to estimate the total cost of each meal. Then estimate the amount each person should pay to split each check evenly.

Number of People: 2

Quantity	Item	Price
2	Large Soda	$0.75 each
1	Cheeseburger	$4.55
1	BLT	$3.25
Estimated Total:		
Estimated Cost Per Person:		

Number of People: 4

Quantity	Item	Price
2	Slice of Pie	$1.45 each
1	Brownie	$1.25
3	Hot Tea	$0.60 each
Estimated Total:		
Estimated Cost Per Person:		

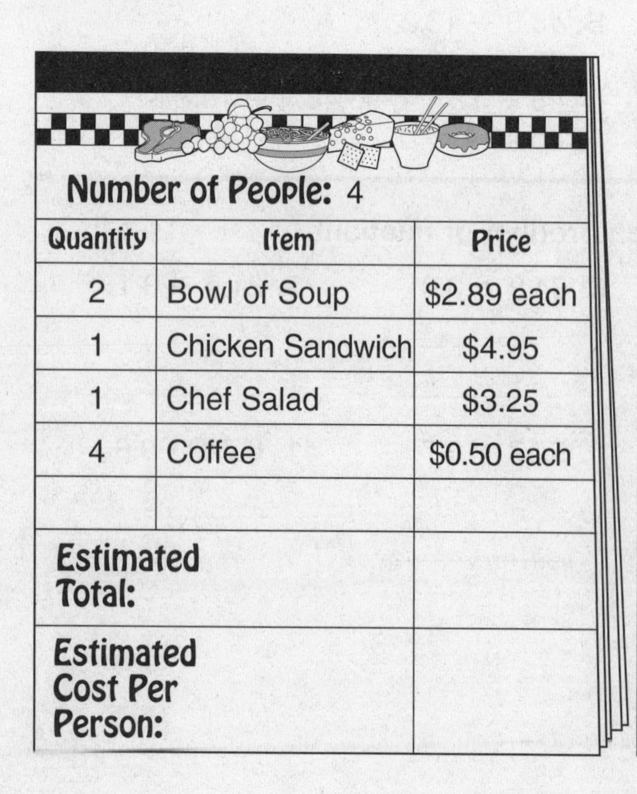

Number of People: 4

Quantity	Item	Price
2	Bowl of Soup	$2.89 each
1	Chicken Sandwich	$4.95
1	Chef Salad	$3.25
4	Coffee	$0.50 each
Estimated Total:		
Estimated Cost Per Person:		

Number of People: 5

Quantity	Item	Price
1	Pizza	$14.95
5	House Salad	$2.85 each
4	Large Soda	$0.75 each
1	Small Soda	$0.55
Estimated Total:		
Estimated Cost Per Person:		

Name ______________________ Date __________ Class __________

LESSON 3-2

Problem Solving

Estimating Decimals

Write the correct answer.

1. Men in Iceland have the highest average life expectancy in the world—76.8 years. The average life expectancy for a man in the United States is 73.1 years. About how much higher is a man's average life expectancy in Iceland? Round your answer to the nearest whole year.

____

2. The average life expectancy for a woman in the United States is 79.1 years. Women in Japan have the highest average life expectancy—3.4 years higher than the United States. Estimate the average life expectancy of women in Japan. Round your answer to the nearest whole year.

____

3. There are about 1.6093 kilometers in one mile. There are 26.2 miles in a marathon race. About how many kilometers are there in a marathon race? Round your answer to the nearest tenths.

____

4. At top speed, a hornet can fly 13.39 miles per hour. About how many hours would it take a hornet to fly 65 miles? Round your answer to the nearest whole number.

____

Circle the letter of the correct answer.

5. The average male human brain weighs 49.7 ounces. The average female human brain weighs 44.6 ounces. What is the difference in their weights?

   A about 95 ounces
   B about 7 ounces
   C about 5 ounces
   D about 3 ounces

6. An official hockey puck is 2.54 centimeters thick. About how thick are two hockey pucks when one is placed on top of the other?

   F about 4 centimeters
   G about 4.2 centimeters
   H about 5 centimeters
   J about 5.2 centimeters

7. Lydia earned $9.75 per hour as a lifeguard last summer. She worked 25 hours a week. About how much did she earn in 8 weeks?

   A about $250.00
   B about $2,000.00
   C about $2,500.00
   D about $200.00

8. Brent mixed 4.5 gallons of blue paint with 1.7 gallons of white paint and 2.4 gallons of red paint to make a light purple paint. About how many gallons of purple paint did he make?

   F about 9 gallons
   G about 8 gallons
   H about 10 gallons
   J about 7 gallons

Name ______________________ Date __________ Class __________

LESSON 3-2

Reading Strategies

Use Context

You estimate to get an approximate answer. Rounding decimals to the nearest whole number is one way to estimate.

Mike's mom bought 3.28 pounds of cheddar cheese. She also bought 2.75 pounds of Swiss cheese. About how many pounds of cheese did she buy?

To round to the nearest whole number, look at the tenths place.

3.**2**8 ⟵ 2 is less than 5; round down to 3.
+2.**7**5 ⟵ 7 is greater than 5; round up to 3.

3.28 + 2.75 rounded to the nearest whole number is:

↓ ↓

3 + 3 = 6 pounds of cheese.

Complete each problem.

1. Which decimal place value do you look at to round to the nearest whole number? ______________________

2. Round 34.67 pounds to the nearest pound. ______________________

3. Round 42.19 pounds to the nearest pound. ______________________

4. Estimate this sum: 42.19 pounds + 34.67 pounds. ______________________

5. Round $54.14 to the nearest dollar. ______________________

6. Round $21.54 to the nearest dollar. ______________________

7. Estimate the difference: $54.14 − $21.54. ______________________

Name ______________________ Date ____________ Class ____________

LESSON **3-2**

Puzzles, Twisters & Teasers

Secret Message

Recently you've noticed that your neighbor Slippery Larry has been extremely forgetful when he leaves his house. He keeps leaving, then returning to get all of the things he forgot. One day you decided to keep track of how far he had gone by measuring how many feet he walked down his driveway before returning to the house. Here's what you found:

1. Keys	6.435
2. Wallet	4.909
3. Jacket	4.899
4. Pants	14.642
5. Watch	12.94
6. Briefcase	4.851
7. Pen	1.491
8. Soccer Ball	13.575
9. Laptop	18.249
10. Suspenders	13.59
11. Rope	1.59
12. Gloves	14.253
13. Hat	18.161

It just so happens that Slippery Larry was on the news later that day. He is actually a criminal communicating to his criminal buddies through a code system. Seems his friends watch how far he travels down his driveway and the distances translate into the following code:

1.6	A	19.0	H	9.4	O	13.0	V
3.6	B	21.2	I	8.8	P	13.8	W
8.4	C	14.8	J	9.6	Q	1.4	X
4.4	D	9.2	K	10.8	R	1.2	Y
5.0	E	2.6	L	18.2	S	5.2	Z
6.8	F	6.4	M	13.6	T		
15.6	G	14.6	N	14.4	U		

Since your measurements are more exact than Slippery Larry's and his friends', you'll have to round each of the distances to get the total distance traveled. Translate the code into letters and fill in the blanks to discover where you can catch Slippery Larry!

____ ____ ____ ____   ____ ____
1 2 2 8   1 2

____ ____ , ____ ____ ____ ____ ____
11 8   13 8 11 4 13

Name ____________________ Date __________ Class __________

LESSON 3-3

Practice A

Adding and Subtracting Decimals

Find each sum or difference.

1. 1.5 + 2.3 ________

2. 6.5 + 1.4 ________

3. 8.9 − 5.1 ________

4. 12.6 − 3.4 ________

5. 8.16 − 7.02 ________

6. 7.25 + 8.75 ________

7. 11.4 + 8.6 ________

8. 16.5 − 4.3 ________

9. 9.55 − 1.2 ________

10. 25.6 + 5.1 ________

11. 8.9 + 3.05 ________

12. 10.64 − 8.5 ________

Circle the letter of the correct answer.

13. If $x = 2.3$, what is the value of the expression $5.4 + x$?

A 3.1
B 7.7
C 7.1
D 3.7

14. If $a = 4.2$, what is the value of the expression $8.7 - a$?

F 12.9
G 4.9
H 4.5
J 12.5

15. If $m = 1.9$, what is the value of the expression $m + 4.2$?

A 2.3
B 2.2
C 6.1
D 7.1

16. If $y = 5.9$, what is the value of the expression $7.2 - y$?

F 1.3
G 1.7
H 13.3
J 13.1

17. Marcus is 1.5 meters tall. His sister, Carol, is 0.1 meter taller than Marcus. Their father is 0.2 meter taller than Carol. How tall is Carol? How tall is their father?

____

18. Jennifer brought $14.75 to the baseball game. She spent $3.45 for a hot dog and soda. How much money does she have left?

____

Name ______________________ Date __________ Class __________

LESSON 3-3

Practice B

Adding and Subtracting Decimals

Find each sum or difference.

1. 8.9 + 2.4 ____________

2. 12.7 − 9.6 ____________

3. 18.35 − 4.16 ____________

4. 7.21 + 11.6 ____________

5. 0.975 + 3.8 ____________

6. 20.66 − 9.1 ____________

7. Tiffany's job requires a lot of driving. How many miles did she travel during the month of February? ____________

Miles Tiffany Traveled

Week	1	2	3	4
Miles	210.05	195.18	150.25	165.30

8. Shelly babysits after school and on the weekends. How much did she earn in all for the month of April? ____________

Shelly's Earnings for April

Week	1	2	3	4
Earnings	$120.50	$180.75	$205.25	$215.50

Evaluate 5.6 – *a* for each value of *a*.

9. $a = 3.7$ ____________

10. $a = 0.5$ ____________

11. $a = 2.8$ ____________

12. $a = 1.42$ ____________

13. $a = 0.16$ ____________

14. $a = 3.75$ ____________

15. Allen bought a box of envelopes for $2.79 and a pack of paper for $4.50. He paid with a $10 bill. How much change should be receive?

____

16. From a bolt of cloth measuring 25.60 yards, Tina cut a 6.8-yard piece and an 11.9-yard piece. How much material is left on the bolt?

____

Name ______________________ Date ____________ Class ____________

LESSON 3-3

Practice C

Adding and Subtracting Decimals

Find each sum or difference.

1. 7.9 + 1.38 ____________

2. 100 − 65.9 ____________

3. 204.965 + 55.8 ____________

4. 63.057 − 18.45 ____________

5. 0.5541 − 0.09 ____________

6. 11.2398 + 8.9 + 2.7 ____________

7. 0.22 + 15.607 + 9.7 ____________

8. 20 − 3.78 − 0.64 ____________

9. 1.9 + 0.25 + 9.4 + 0.5 ____________

Evaluate each expression.

10. 39.702 – a for $a = 0.9$ ____________

11. $x + 1.064$ for $x = 28.5$ ____________

12. 50.02 – p for $p = 0.99$ ____________

13. $w - 7.08$ for $w = 100$ ____________

14. $37.62 + t$ for $t = 8.084$ ____________

15. $u + 2.7$ for $u = 0.5046$ ____________

Write the missing digit in each problem.

16.
```
   7.089
+  2.□13
   9.502
```
______

17.
```
  16.594
−  □.175
  11.419
```
______

18.
```
   6.2□67
+  9.75
  15.9867
```
______

19.
```
   0.6□9
−  0.458
   0.221
```
______

20.
```
  238.793
+  19.5□2
  258.305
```
______

21.
```
  100.□5
−  19.99
   80.46
```
______

22. Italian Delight sells three sizes of pizzas at different prices. If you buy all three pizzas, it costs a total of $46.24. A medium pizza costs $15.75, and a large costs $17.50. How much does a small pizza cost?

____

23. Brent has three sheets of plywood that are each 6.85 feet long. He cut a 3.4-foot piece from one sheet and a 0.5-foot piece from the other. How many feet of plywood does he have left in all?

____

Name ______________________ Date ____________ Class ____________

LESSON **3-3**

Reteach

Adding and Subtracting Decimals

You can use a place-value chart to help you add and subtract decimals.

Add 1.4 and 0.9.

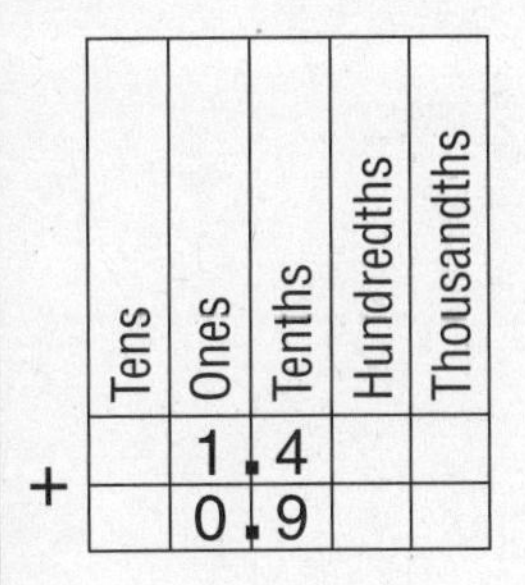

	Tens	Ones	Tenths	Hundredths	Thousandths
		1.	4		
+		0.	9		

So, 1.4 + 0.9 = 2.3.

Subtract 2.4 from 3.1.

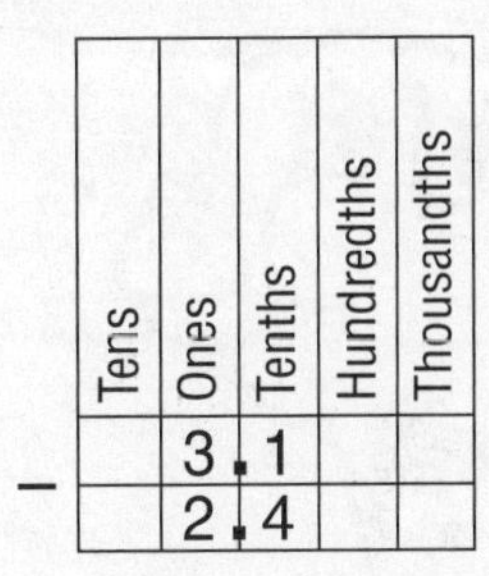

	Tens	Ones	Tenths	Hundredths	Thousandths
		3.	1		
–		2.	4		

So, 3.1 – 2.4 = 0.7.

Find each sum or difference.

1.

	Tens	Ones	Tenths	Hundredths	Thousandths
		2.	6		
+		1.	1	5	

____

2.

	Tens	Ones	Tenths	Hundredths	Thousandths
		2.	5	3	
–		1.	7		

____

3. 4.3 + 1.4

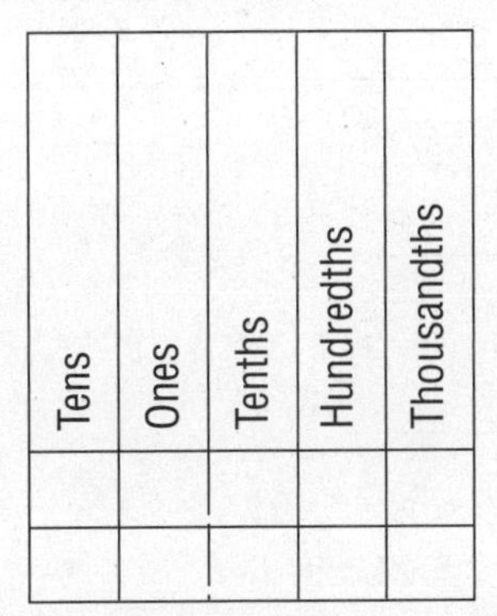

Tens	Ones	Tenths	Hundredths	Thousandths

____

4. 14.4 – 3.8

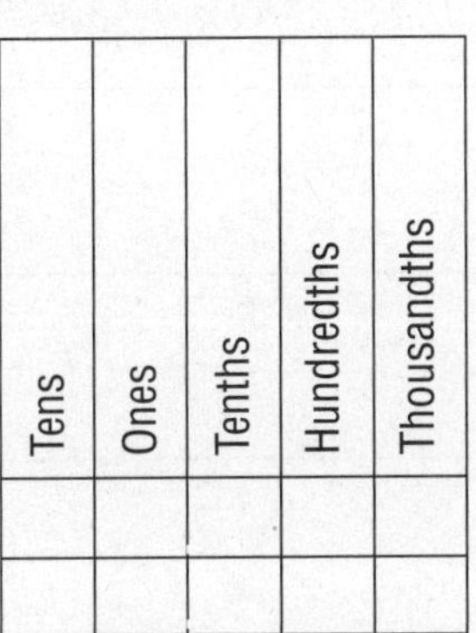

Tens	Ones	Tenths	Hundredths	Thousandths

____

5. 7.3 + 8.5

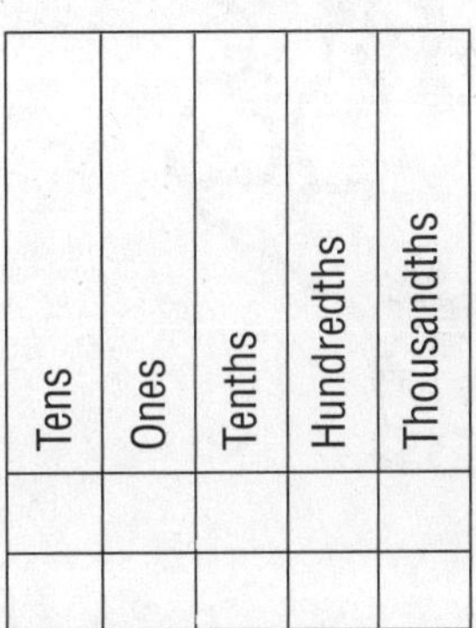

Tens	Ones	Tenths	Hundredths	Thousandths

____

6. 12.34 – 6.9

Tens	Ones	Tenths	Hundredths	Thousandths

____

7. 6.3 – 2.5

____

8. 20.65 + 13.24

____

9. 8.9 – 1.95

____

10. 3.42 + 5.25

____

Name ______________________ Date __________ Class __________

LESSON 3-3

Challenge

A Penny Saved Is a Penny Earned

Next to each bank, describe three different coin combinations that equal the amount of money it holds. For each combination, use at least one quarter, one dime, one nickel, and one penny.

1.

____

____

____

2.

____

____

____

3.

____

____

____

4.

____

____

____

Name ______________________ Date __________ Class __________

LESSON 3-3

Problem Solving

Adding and Subtracting Decimals

Use the table to answer the questions.

Busiest Ports in the United States

Port	Imports Per Year (millions of tons)	Exports Per Year (millions of tons)
South Louisiana, LA	30.6	57.42
Houston, TX	75.12	33.43
New York, NY & NJ	53.52	8.03
New Orleans, LA	26.38	21.73
Corpus Christi, TX	52.6	7.64

1. How many more tons of imports than exports does the Port of New Orleans handle each year?

   ____

2. How many tons of imports and exports are shipped through the port of Houston, Texas, each year in all?

   ____

Circle the letter of the correct answer.

3. Which port ships 0.39 more tons of exports each year than the port at Corpus Christi, Texas?
   - **A** Houston
   - **B** NY & NJ
   - **C** New Orleans
   - **D** South Louisiana

4. What is the difference between the imports and exports shipped in and out of Corpus Christi's port each year?
   - **F** 45.04 million tons
   - **G** 44.94 million tons
   - **H** 44.96 million tons
   - **J** 44.06 million tons

5. What is the total amount of imports shipped into the nation's 5 busiest ports each year?
   - **A** 238.22 million tons
   - **B** 366.47 million tons
   - **C** 128.25 million tons
   - **D** 109.97 million tons

6. What is the total amount of exports shipped out of the nation's 5 busiest ports each year?
   - **F** 366.47 million tons
   - **G** 128.25 million tons
   - **H** 109.97 million tons
   - **J** 238.22 million tons

Name ______________________ Date ___________ Class ___________

LESSON 3-3

Reading Strategies

Use an Organizer

Writing decimals in a place-value grid helps you line up decimal points to add or subtract decimals.

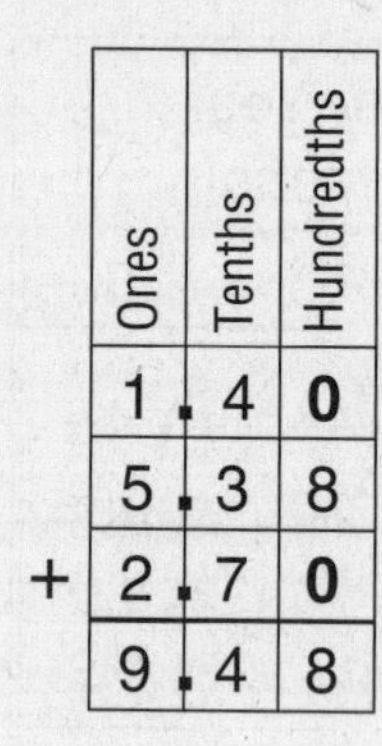

	Ones	Tenths	Hundredths
	1.	4	**0**
	5.	3	8
+	2.	7	**0**
	9.	4	8

$$\begin{array}{r} 1.4\mathbf{0} \\ 5.38 \\ +\,2.7\mathbf{0} \\ \hline 9.48 \end{array}$$

Add zeros as place holders.

Place decimal point in answer.

$$\begin{array}{r} 28.05 \\ -\,6.3\mathbf{0} \\ \hline 21.75 \end{array}$$

	Tens	Ones	Tenths	Hundredths
	2	8.	0	5
−		6.	3	**0**
	2	1.	7	5

1. How does the place-value grid help you add or subtract?

____

____

2. Place these numbers on the place-value grid below: 3.25, 1.06, 2.9.

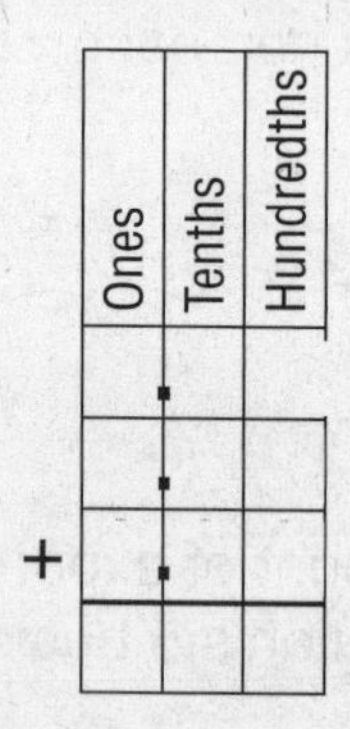

3. Add the numbers on the place-value grid. What is the sum?

____

4. Place this problem on the place-value grid below: 23.82 − 7.2.

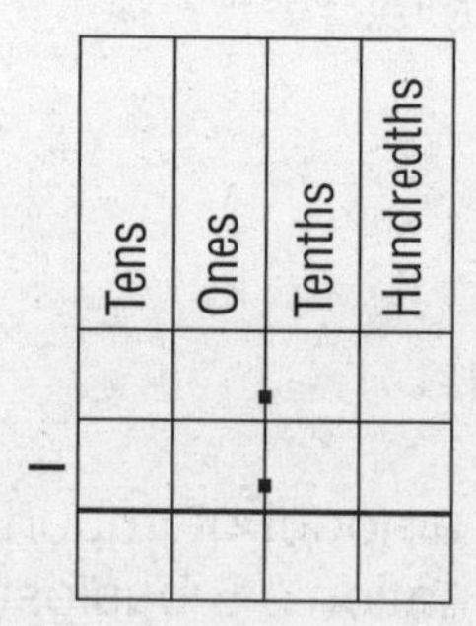

5. Subtract the numbers on the place-value grid. What is the difference?

____

6. For which numbers did you add zero as a place holder?

____

Name ______________________ Date ___________ Class ___________

LESSON **3-3**

Puzzles, Twisters & Teasers

Crazy Pete's Letter Shop

What's the quickest way to double your money?

To solve this riddle, you need to buy your letters from Crazy Pete.

Right now, you know three letters from the first word: **L, F, O.** You also know three letters from the second word: **F, A, H.**

To use those six letters in the riddle, you need to buy them from Crazy Pete.

You have $15.00 to spend on each word, but you know you need to buy two more letters to complete the riddle—one for each word.

Crazy Pete tells you a secret: "You have exactly enough money left after buying the first three letters of each word to purchase the missing letter."

Using the price table at right, add the cost of the three letters you need to buy for the first word. Subtract this total from $15.00 to find the cost of the missing letter.

Then do it again for the second word.

A	$1.50
B	$5.43
C	$6.08
D	$4.13
E	$1.50
F	$6.08
G	$4.79
H	$2.83
I	$1.50
J	$6.08
K	$4.79
L	$4.59
M	$5.43
N	$3.29
O	$0.20
P	$4.79
Q	$7.87
R	$3.29
S	$3.29
T	$3.29
U	$1.50
V	$4.79
W	$5.43
X	$7.87
Y	$5.43
Z	$7.87

First Word Cost (letters **L, F, O**): ______________

Remaining money: ______________

Missing Letter: ______________

Second Word Cost (letters **F, A, H**): ______________

Remaining money: ______________

Final Letter: ______________

Unscramble the letters of each word to solve the riddle!

___ ___ ___ ___ it in

___ ___ ___ ___!

Name ______________________ Date ____________ Class ____________

LESSON 3-4

Practice A

Scientific Notation

Find each product.

1. 267 • 100

____

2. 38.1 • 100

____

3. 1.92 • 100

____

Circle the letter of the correct answer.

4. Which of the following shows 85,000 written in scientific notation?

A $8.5 \cdot 10^3$

B $8.5 \cdot 10^4$

C $8.5 \cdot 10^5$

D $8.5 \cdot 10^6$

5. Which of the following shows $3.67 \cdot 10^5$ written in standard form?

F 3,670

G 36,700

H 367,000

J 3,670,000

Fill in the blanks to make each equation true.

6. $1{,}200 = 1.2 \cdot 10^{—}$

____

7. $25{,}000 = 2.5 \cdot 10^{—}$

____

8. $580 = 5.8 \cdot 10^{—}$

____

9. $470{,}000 =$ ____ $\cdot 10^5$

____

10. $6{,}580 =$ ____ $\cdot 10^3$

____

11. $8{,}900{,}000 =$ ____ $\cdot 10^6$

____

Write each number in standard form.

12. $3.4 \cdot 10^2$

____

13. $7.9 \cdot 10^4$

____

14. $1.75 \cdot 10^3$

____

15. $1.24 \cdot 10^5$

____

16. $9.6 \cdot 10^5$

____

17. $1.28 \cdot 10^6$

____

18. African elephants are the largest land mammals. The average African elephant weighs 11,000 pounds. Write this weight in scientific notation.

____

Name ______________________ Date __________ Class __________

LESSON 3-4

Practice B

Scientific Notation

Find each product.

1. 345 • 100 ____________

2. 65.2 • 100 ____________

3. 1.84 • 1,000 ____________

Write each number in scientific notation.

4. 16,700 ____________

5. 4,680 ____________

6. 58,340,000 ____________

Write each number in standard form.

7. $3.25 \cdot 10^4$ ____________

8. $7.08 \cdot 10^6$ ____________

9. $1.209 \cdot 10^7$ ____________

10. $6.8 \cdot 10^8$ ____________

11. $0.51 \cdot 10^5$ ____________

12. $0.006 \cdot 10^3$ ____________

Identify the answer choice that is *not* equal to the given number.

13. 356,000

A 300,000 + 56,000

B $3.56 \cdot 10^5$

C $3.56 \cdot 10^4$

14. $1.28 \cdot 10^6$

A 100,000 + 28,000

B 1,280,000

C $12.8 \cdot 10^5$

15. 1,659,000

A 1,600,000 + 59,000

B $1.659 \cdot 10^6$

C $16.59 \cdot 10^6$

16. $0.074 \cdot 10^3$

A 70.0 + 4.0

B $7.4 \cdot 10^5$

C $7.4 \cdot 10^1$

17. In 2000, the population of Pennsylvania was 12,281,054. Round this figure to the nearest hundred thousand. Then write that number in scientific notation.

____

18. In 2000, the population of North Carolina was about $8.05 \cdot 10^6$, and the population of South Carolina was about $4.01 \cdot 10^6$. Write the combined populations of these two states in standard form.

____

Name ______________________ Date __________ Class __________

LESSON 3-4

Practice C

Scientific Notation

Find each product.

1. 1.67 x 1,000 ______

2. 93.6 x 100 ______

3. 3.55 x 10,000 ______

Write each number in scientific notation.

4. 6,389,000 ______

5. 105,200,000 ______

6. 152 million ______

Write each number in standard form.

7. $1.5089 \cdot 10^4$ ______

8. $2.516 \cdot 10^8$ ______

9. $1.7711 \cdot 10^7$ ______

10. $3.9604 \cdot 10^6$ ______

11. $0.284 \cdot 10^4$ ______

12. $0.0869 \cdot 10^2$ ______

Write each measurement using scientific notation.

13. 250 km = ______ m

14. 0.065 kg = ______ g

15. 89 L = ______ mL

16. 1,540 km = ______ cm

17. 0.73 m = ______ mm

18. 10,240 kg = ______ g

19. In a recent count, 147,171,000 people in the United States owned cars. In the same year, $4.268 \cdot 10^7$ people in Japan owned cars. In which country did more people own cars? How many more?

______

______

20. On average, about $1.1 \cdot 10^9$ passengers use the New York City subway system each year. About 1,170,000,000 passengers use the Paris subway each year. How many passengers use those two subways each year?

______

______

Name ______________________ Date __________ Class __________

LESSON 3-4

Reteach

Scientific Notation

Scientific notation expresses a large number as the product of a number between one and ten and a power of ten.

To write 3,400 in scientific notation, move the decimal point to the left until the number falls between 1 and 10.

3,400 $1 < 3 < 10$, so move the decimal point 3 places to the left.

$3{,}400 = 3.4 \bullet 10^3$ (3 2 1) The number of times you move the decimal point left is the power of ten.

Express each number in scientific notation.

1. 175,000 ____________

2. 298 ____________

3. 5,764 ____________

4. 83 ____________

5. 40,300 ____________

6. 2,000,000 ____________

7. 51,010 ____________

8. 190,025 ____________

You can express numbers written in scientific notation in standard form.

The power of ten tells you how many places to move the decimal point to the right.

$3.2 \bullet 10^4 = 32{,}000$ To write $3.2 \bullet 10^4$ in standard form, move the decimal point 4 places to the right.

Write each number in standard form.

9. $5.62 \bullet 10^3$ ____________

10. $7.238 \bullet 10^2$ ____________

11. $9.9 \bullet 10^5$ ____________

12. $6.53 \bullet 10^1$ ____________

13. $5.36 \bullet 10^4$ ____________

14. $2.4 \bullet 10^2$ ____________

15. $4.35 \bullet 10^3$ ____________

16. $8 \bullet 10^5$ ____________

17. $1 \bullet 10^4$ ____________

18. $2.03 \bullet 10^3$ ____________

19. $1.12 \bullet 10^2$ ____________

20. $3.002 \bullet 10^6$ ____________

Name ______________________ Date ____________ Class ____________

LESSON 3-4

Challenge

The Solar System

Write the average distance from the Sun in standard form. Then use the distances to label our solar system shown below.

		Average Distance From the Sun (mi)	
		Scientific Notation	**Standard Form**
1.	Earth	$9.29 \cdot 10^7$	
2.	Jupiter	$4.836 \cdot 10^8$	
3.	Mars	$1.416 \cdot 10^8$	
4.	Mercury	$3.6 \cdot 10^7$	
5.	Neptune	$2.794 \cdot 10^9$	
6.	Pluto	$3.675 \cdot 10^9$	
7.	Saturn	$8.87 \cdot 10^8$	
8.	Uranus	$1.784 \cdot 10^9$	
9.	Venus	$6.72 \cdot 10^7$	

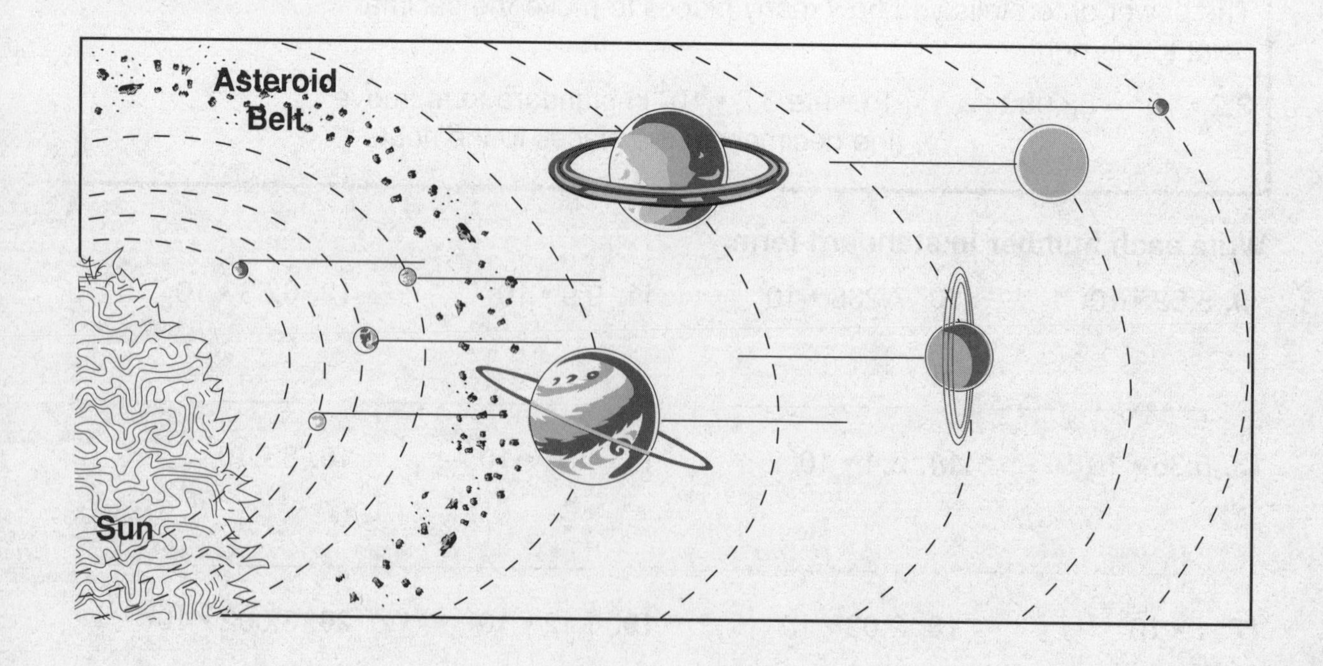

Name ______________________ Date __________ Class __________

LESSON 3-4

Problem Solving

Scientific Notation

Write the correct answer.

1. The closest comet to approach Earth was called Lexell. On July 1, 1770, Lexell was observed about 874,200 miles from Earth's surface. Write this distance in scientific notation.

____

2. Scientists estimate that it would take $1.4 \bullet 10^{10}$ years for light from the edge of our universe to reach Earth. How many years is that written in standard form?

____

3. In the United States, about 229,000,000 people speak English. About 18,000,000 people speak English in Canada. Write in scientific notation the total number of English speaking people in the United States and Canada.

____

4. South Africa is the top gold-producing country in the world. Each year it produces $4.688 \bullet 10^8$ tons of gold! Written in standard form, how many tons of gold does South African produce each year?

____

Circle the letter of the correct answer.

5. About $3.012 \bullet 10^6$ people visit Yellowstone National Park each year. What is that figure written in standard form?

   A 30,120,000 people
   B 3,012,000 people
   C 301,200 people
   D 30,120 people

6. In 2000, farmers in Iowa grew 1,740,000 bushels of corn. What is this amount written in scientific notation?

   F $1.7 \bullet 10^5$
   G $1.74 \bullet 10^5$
   H $1.74 \bullet 10^6$
   J $1.74 \bullet 10^7$

7. The temperature at the core of the Sun reaches 27,720,000°F. What is this temperature written in scientific notation?

   A $2.7 \bullet 10^7$
   B $2.72 \bullet 10^7$
   C $2.772 \bullet 10^6$
   D $2.772 \bullet 10^7$

8. Your body is constantly producing red blood cells—about $1.73 \bullet 10^{11}$ cells a day. How many blood cells is that written in standard form?

   F 173,000,000 cells
   G 17,300,000,000 cells
   H 173,000,000,000 cells
   J 1,730,000,000,000 cells

Name ______________________ Date __________ Class __________

LESSON 3-4

Reading Strategies
Use a Graphic Organizer

This chart helps you see the ways large numbers can be written.

Ways to Write Large Numbers

Scientific Notation

- A number between 1 and 10 multiplied by a power of 10
  2.5×10^6
  7×10^5

Standard Form	**Words and Symbols**
• All place values are shown. 2,500,000 700,000	• Use numbers and words. 2 million, 500 thousand; 2.5 million 700 thousand

Use the graphic organizer to answer Exercises 1–2.

1. Which way to write large numbers shows every place value?

____

2. Which way to write large numbers uses a power of 10?

____

Identify how each large number is written. Write "scientific notation," "standard form," or "words and symbols".

3. 8,296,000

____

4. 3.6 million

____

5. 2.9×10^5

____

Name ______________________ Date ____________ Class ____________

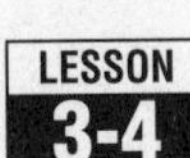

LESSON 3-4

Puzzles, Twisters & Teasers

The Scientific Two-Step

There are two simple steps to do the Scientific Two-Step and answer the riddle:

1. For each number in scientific notation in the left column, find the same number in standard notation in the key on the right.
2. Fill in the letter associated with the standard notation answer to solve the riddle.

1. 2.458×10^1
2. 1.276×10^8
3. 2.458×10^2
4. 1.276×10^6
5. 2.458×10^4
6. 2.458×10^9
7. 1.276×10^7
8. 2.458×10^3
9. 2.458×10^5
10. 1.276×10^3
11. 2.458×10^7
12. 1.276×10^4
13. 1.276×10^2

Key:

12,760	E
127,600,000	O
24,580,000	D
245.8	M
24,580	A
2,458	F
12,760,000	Y
1,276	L
24.58	C
245,800	O
1,276,000	P
2,458,000,000	N
127.6	D

Why did the people working at the blanket factory lose their jobs?

The ___ ___ ___ ___ ___ ___ ___
1 2 3 4 5 6 7

___ ___ ___ ___ ___ ___
8 9 10 11 12 13

Name ______________________ Date __________ Class __________

LESSON 3-5

Practice A

Multiplying Decimals

Find each product.

1. 0.4×0.2 ________

2. 0.3×0.4 ________

3. 1.2×0.5 ________

4. 1.1×0.9 ________

5. 2.5×0.5 ________

6. 6.0×0.7 ________

7. $0.4 \cdot 0.5$ ________

8. $1.2 \cdot 1.5$ ________

9. $1.7 \cdot 0.3$ ________

10. $6.7 \cdot 0.4$ ________

11. $9.6 \cdot 0.2$ ________

12. $0.8 \cdot 0.8$ ________

Evaluate $2x$ for each value of x.

13. $x = 0.1$ ________

14. $x = 0.5$ ________

15. $x = 0.9$ ________

16. $x = 1.2$ ________

17. $x = 1.7$ ________

18. $x = 2.4$ ________

19. Each box can hold 2.5 pounds of apples. How many pounds can 3 boxes hold?

____

20. Each pie costs $5.60. How much will it cost to buy 2 pies?

____

Name ______________________ Date __________ Class __________

LESSON 3-5

Practice B

Multiplying Decimals

Find each product.

1. 0.7×0.3 ________

2. 0.05×0.4 ________

3. 8.0×0.02 ________

4. 3.5×0.2 ________

5. 12.1×0.01 ________

6. 9.0×0.9 ________

7. $0.04 \bullet 0.58$ ________

8. $2.15 \bullet 1.5$ ________

9. $1.73 \bullet 0.8$ ________

10. $6.017 \bullet 2.0$ ________

11. $3.96 \bullet 0.4$ ________

12. $0.7 \bullet 0.009$ ________

Evaluate 8*x* for each value of *x*.

13. $x = 0.5$ ________

14. $x = 2.3$ ________

15. $x = 0.74$ ________

16. $x = 3.12$ ________

17. $x = 0.587$ ________

18. $x = 14.08$ ________

19. The average mail carrier walks 4.8 kilometers in a workday. How far do most mail carriers walk in a 6-day week? There are 27 working days in July, so how far will a mail carrier walk in July?

____

____

20. A deli charges $3.45 for a pound of turkey. If Tim wants to purchase 2.4 pounds, how much will it cost?

____

____

Name ______________________ Date ____________ Class ____________

LESSON 3-5

Practice C

Multiplying Decimals

Find each product.

1. 9.86×0.3 ______

2. 12.01×0.46 ______

3. 7.05×0.03 ______

4. 11.65 • 0.23 ______

5. 24.54 • 0.037 ______

6. 11.405 • 2.91 ______

7. 0.058 • 0.129 ______

8. 29.864 • 5.13 ______

9. 100.86 • 0.004 ______

Evaluate 17*x* for each value of *x*.

10. $x = 1.9$ ______

11. $x = 0.005$ ______

12. $x = 6.307$ ______

13. $x = 11.215$ ______

14. $x = 2.059$ ______

15. $x = 75.844$ ______

Evaluate.

16. $2.97n$ for $n = 1.8$ ______

17. $12^2 + 1.9c$ for $c = 3.7$ ______

18. $7^3 - 2x$ for $x = 0.54$ ______

19. $1.6t + 3.056$ for $t = 2.09$ ______

20. One year on Mercury is equal to 87.97 Earth days. One year on Pluto is three times the length of one Mercury year minus 16.21 days. How long is one year on Pluto?

______

21. One year on Earth is equal to 365.30 days. One year on Mars is twice the length of one Earth year minus 43.6 days. How long is one year on Mars?

______

Name ______________________ Date ___________ Class ___________

LESSON 3-5

Reteach

Multiplying Decimals

You can use a model to help you multiply a decimal by a whole number.

Find the product of 0.12 and 4, using a 10 by 10 grid.

Shade 4 groups of 12 squares. Count the number of shaded squares. Since you have shaded 48 of the 100 squares, $0.12 \bullet 4 = 0.48$.

Find each product.

1. $0.23 \bullet 3$ ________
2. $0.41 \bullet 2$ ________
3. $0.011 \bullet 5$ ________
4. $0.32 \bullet 2$ ________
5. $0.15 \bullet 3$ ________
6. $0.42 \bullet 2$ ________
7. $0.04 \bullet 8$ ________
8. $0.22 \bullet 4$ ________

You can also use a model to help you multiply a decimal by a decimal.

Find the product of 0.4 and 0.6.

$0.4 \bullet 0.6 = 0.24$

Find each product.

9. $0.2 \bullet 0.8$ ________
10. $0.7 \bullet 0.9$ ________
11. $0.5 \bullet 0.5$ ________
12. $0.3 \bullet 0.6$ ________
13. $0.5 \bullet 0.2$ ________
14. $0.4 \bullet 0.4$ ________
15. $0.1 \bullet 0.9$ ________
16. $0.4 \bullet 0.7$ ________

Name ______________________ Date ____________ Class ____________

LESSON 3-5

Challenge

Decimal Growth

Use the growth rate for each plant below to find how much it will grow in 1 week.

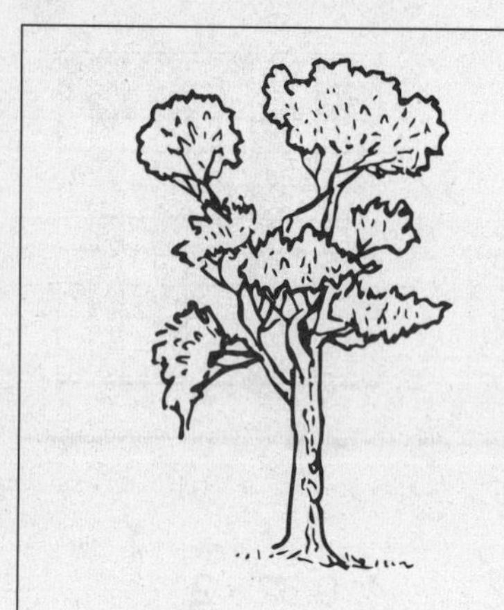		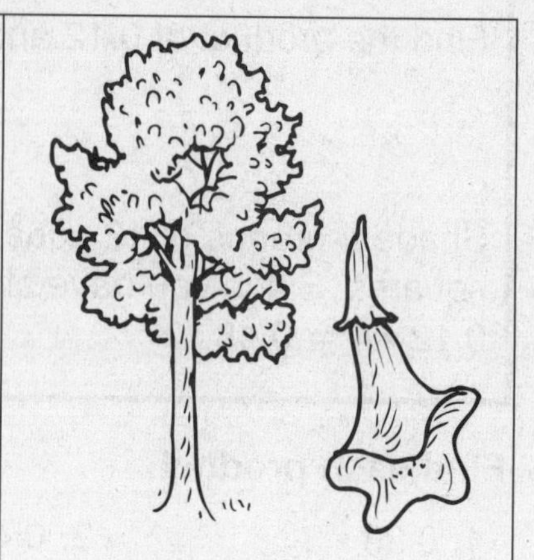
Eucalyptus Tree **Growth Rate:** **2.5 cm per day** ____________	**Bristlecone Pine Tree** **Growth Rate:** **0.009 mm per day** ____________	**Trumpet Tree** **Growth Rate:** **0.28 in. per day** ____________

Use the growth rate for each plant below to find how much it will grow in 0.25 day.

	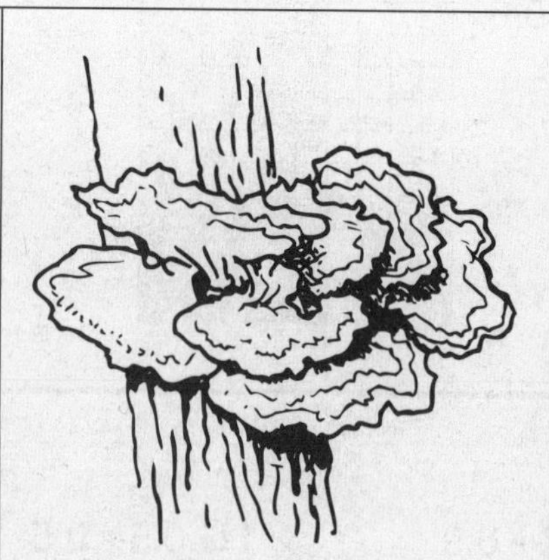	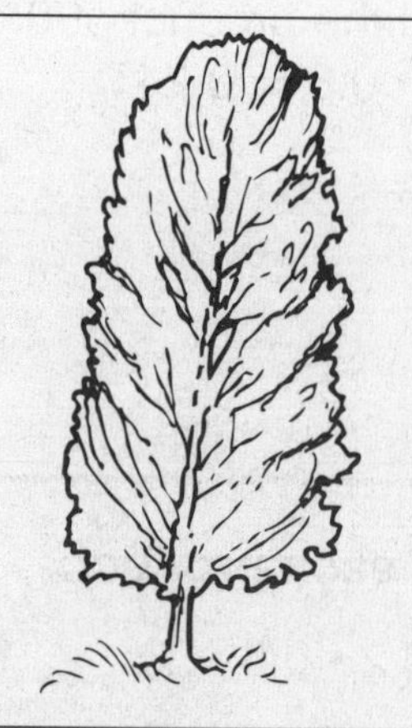
Oak Tree **Growth Rate:** **1.4 mm per day** ____________	**Lichens** **Growth Rate:** **0.0025 mm per day** ____________	**Poplar Tree** **Growth Rate:** **0.118 in. per day** ____________

Name ________________________ Date __________ Class __________

LESSON 3-5

Problem Solving

Multiplying Decimals

Use the table to answer the questions.

United States Minimum Wage

Year	Hourly Rate
1940	$0.30
1950	$0.75
1960	$1.00
1970	$1.60
1980	$3.10
1990	$3.80
2000	$5.15

1. At the minimum wage, how much did a person earn for a 40-hour workweek in 1950?

____

2. At the minimum wage, how much did a person earn for working 25 hours in 1970?

____

3. If you had a minimum-wage job in 1990, and worked 15 hours a week, how much would you have earned each week?

____

4. About how many times higher was the minimum wage in 1960 than in 1940?

____

Circle the letter for the correct answer.

5. Ted's grandfather had a minimum-wage job in 1940. He worked 40 hours a week for the entire year. How much did Ted's grandfather earn in 1940?

   A $12.00
   B $624.00
   C $642.00
   D $6,240.00

6. Marci's mother had a minimum-wage job in 1980. She worked 12 hours a week. How much did Marci's mother earn each week?

   F $3.72
   G $37.00
   H $37.10
   J $37.20

7. Having one dollar in 1960 is equivalent to having $5.82 today. If you worked 40 hours a week in 1960 at minimum wage, how much would your weekly earnings be worth today?

   A $40.00
   B $5.82
   C $232.80
   D $2,328.00

8. In 2000, Cindy had a part-time job at a florist, where she earned minimum wage. She worked 18 hours each week for the whole year. How much did she earn from this job in 2000?

   F $927.00
   G $4,820.40
   H $10,712.00
   J $2,142.40

Name ______________________ Date ____________ Class ____________

LESSON 3-5

Reading Strategies

Use a Visual Tool

Each grid shows 0.15 shaded.

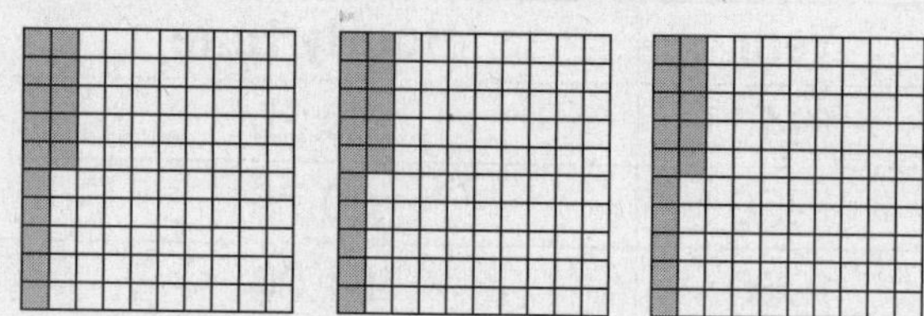

You can add the decimals to find how much of the grids are shaded. ⟶ $0.15 + 0.15 + 0.15 = 0.45$

You can multiply 0.15 by 3. ⟶

$$\begin{array}{r} \overset{1}{0.15} \\ \times \quad 3 \\ \hline 0.45 \end{array}$$

Use these grids to complete the problems below.

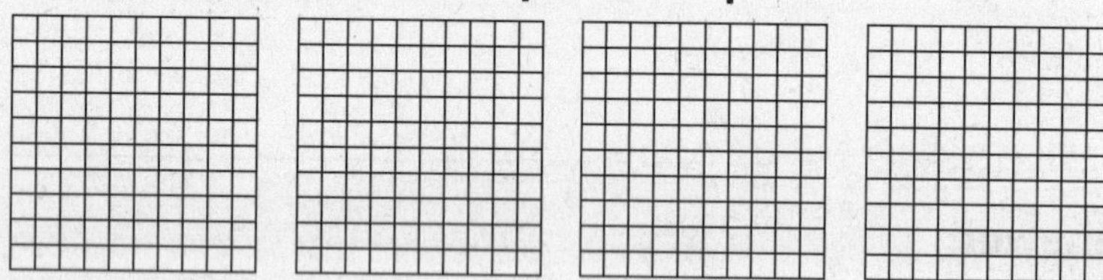

1. Shade 0.23 in each of the 4 grids.
2. Write an addition problem for the shaded grids.

   ____

3. Find the sum of your addition problem.

   ____

4. Write a multiplication problem for your shaded picture.

   ____

5. Find the product of your multiplication problem.

   ____

Name ______________________ Date __________ Class __________

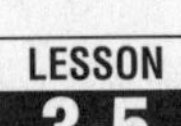

Puzzles, Twisters & Teasers

Maria's Twice Cut Cake

Today is Maria's birthday. Maria's coworkers, Bob, Makiko, José, and Edna, bought her a cake. Unfortunately, Maria and her coworkers are dieting, so none of them wanted a whole piece of cake. They each wanted a fraction of a piece. Maria wanted 0.2 of a piece, Bob wanted 0.16 of a piece, Makiko wanted 0.32 of
a piece, José wanted 0.225 of a piece and Edna wanted just 0.095 of a piece. Because Maria was not a mathematician, she misunderstood her coworkers—she thought that they wanted the entire cake divided into 5 pieces. So she cut the cake as shown.

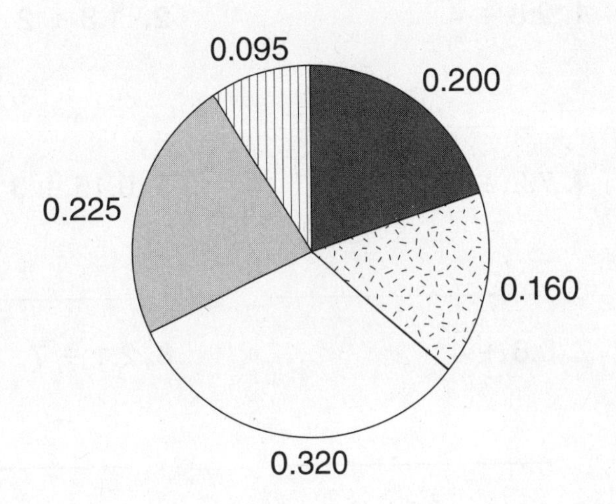

When the cake was cut, the coworkers received the wrong pieces. Maria got Bob's piece, Bob got Makiko's piece, Makiko got José's piece, José got Edna's piece, and Edna got Maria's piece.

Once they received their pieces, each coworker assumed they received a full piece of cake so each of them cut off the amount they had originally requested.

For example, Maria received Bob's 0.16 of the cake—so she ate her 0.2 piece of that. Mathematically she ate 0.2 x 0.16 = 0.032 piece of the cake. The question is: Who ate the most cake? List how much of the entire cake each of the five coworkers ate, then decide who ate the most cake.

Maria has 0.032 of a piece.

Bob has ______________

Makiko has ______________

José has ______________

Edna has ______________

So, who ate the most cake? ______________

Name ______________________ Date __________ Class __________

LESSON 3-6

Practice A

Dividing Decimals by Whole Numbers

Find each quotient.

1. $2.8 \div 4$ ________

2. $1.8 \div 2$ ________

3. $3.6 \div 6$ ________

4. $7.2 \div 9$ ________

5. $0.15 \div 3$ ________

6. $4.8 \div 8$ ________

7. $0.8 \div 4$ ________

8. $2.1 \div 7$ ________

9. $0.32 \div 4$ ________

10. $5.4 \div 9$ ________

11. $3.5 \div 5$ ________

12. $0.2 \div 2$ ________

Evaluate $2.4 \div x$ for each given value of x.

13. $x = 8$ ________

14. $x = 2$ ________

15. $x = 3$ ________

16. $x = 4$ ________

17. $x = 6$ ________

18. $x = 12$ ________

19. A six-pack of orange soda costs $4.20. How much does each can in the pack cost?

____

20. It rained 2.7 inches in July and 2.1 inches in August. What was the average rainfall for those two months?

____

Name ______________________ Date __________ Class __________

LESSON 3-6

Practice B

Dividing Decimals by Whole Numbers

Find each quotient.

1. $0.81 \div 9$ ______________

2. $1.84 \div 4$ ______________

3. $7.2 \div 6$ ______________

4. $13.6 \div 8$ ______________

5. $4.55 \div 5$ ______________

6. $29.6 \div 8$ ______________

7. $15.57 \div 9$ ______________

8. $0.144 \div 12$ ______________

9. $97.5 \div 3$ ______________

10. $0.0025 \div 5$ ______________

11. $2.84 \div 8$ ______________

12. $18.9 \div 3$ ______________

Evaluate $2.094 \div x$ for each given value of x.

13. $x = 2$ ______________

14. $x = 4$ ______________

15. $x = 12$ ______________

16. $x = 20$ ______________

17. $x = 15$ ______________

18. $x = 30$ ______________

19. There are three grizzly bears in the city zoo. Yogi weighs 400.5 pounds, Winnie weighs 560.35 pounds, and Nyla weighs 618.29 pounds. What is the average weight of the three bears?

____

20. The bill for dinner came to $75.48. The four friends decided to leave a $15.00 tip. If they shared the bill equally, how much will they each pay?

____

Name ______________________ Date __________ Class __________

LESSON 3-6

Practice C

Dividing Decimals by Whole Numbers

Find each quotient.

1. $2.36 \div 8$

2. $0.1488 \div 3$

3. $72.654 \div 6$

4. $8.523 \div 9$

5. $115.8 \div 12$

6. $0.952 \div 17$

7. $46.545 \div 29$

8. $14.795 \div 55$

9. $0.2808 \div 75$

Evaluate $x \div 6$ for each value of x.

10. $x = 4.8$

11. $x = 0.54$

12. $x = 0.024$

13. $x = 1.08$

14. $x = 0.42$

15. $x = 0.0012$

Evaluate.

16. $n \div 19$ for $n = 28.5$

17. $(6^2 + 1.35) \div c$ for $c = 5$

18. $4^3 - (0.81 \div x)$ for $x = 9$

19. $3.5t \div 4$ for $t = 19.36$

20. As of 2000, there were 281.42 million people in the United States. If the same number of people lived in each of the 50 states, what would have been the population of each state in 2000?

21. In a gymnastics competition, Kim scored 9.4, 9.7, 9.9, and 9.8. Tamara scored 9.5, 9.2, 9.7, and 9.6. Who had the highest average score?

Name ______________________ Date __________ Class __________

LESSON 3-6

Reteach

Dividing Decimals by Whole Numbers

You can use decimal grids to help you divide decimals by whole numbers.

To divide 0.35 by 7, first shade in a decimal grid to show thirty-five hundredths.

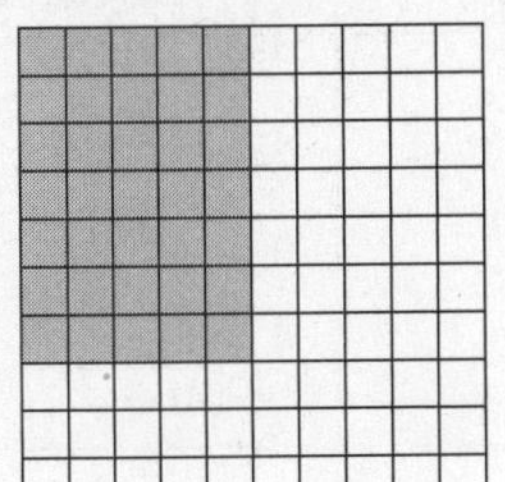

0.35 ÷ 7 means "divide 0.35 into 7 equal groups." Show this on the decimal grid.

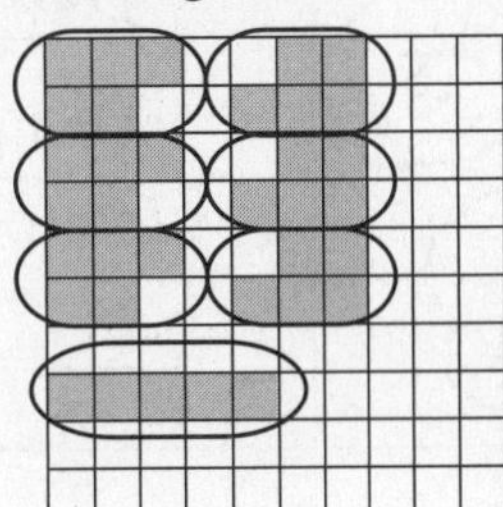

The number of units in each group is the quotient.

So, 0.35 ÷ 7 = 0.05.

Use decimal grids to find each quotient.

1. 0.24 ÷ 4

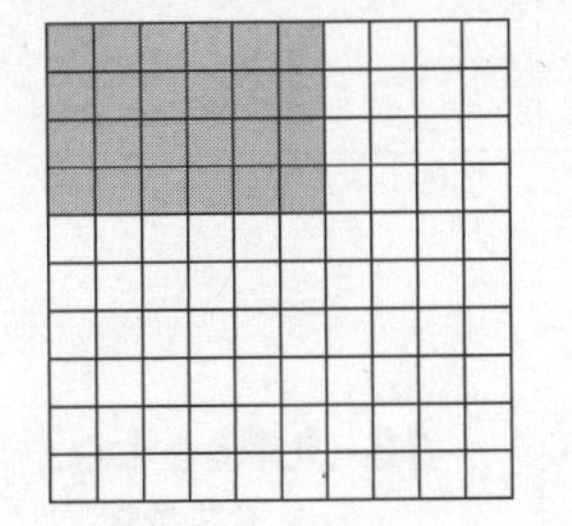

____

2. 0.48 ÷ 12

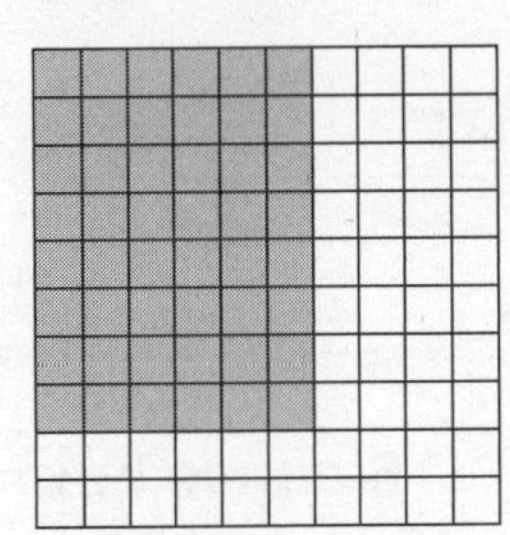

____

3. 0.50 ÷ 10

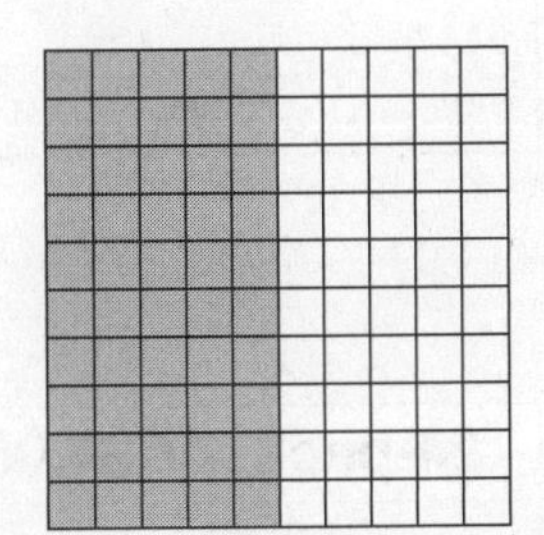

____

4. 0.98 ÷ 7

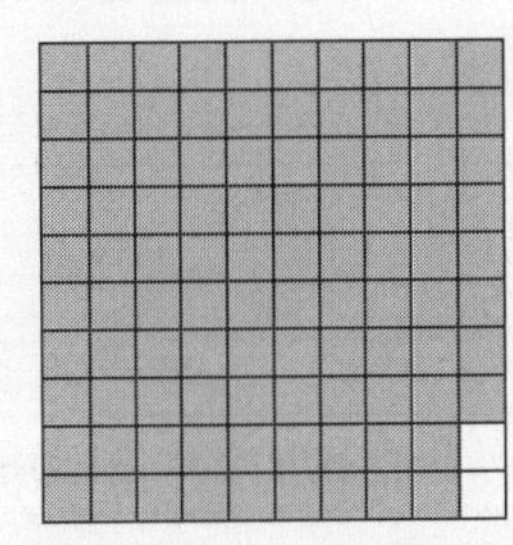

____

5. 0.6 ÷ 5 __________

6. 0.78 ÷ 6 __________

7. 0.99 ÷ 11 __________

8. 0.32 ÷ 4 __________

Name ______________________ Date __________ Class __________

LESSON 3-6

Challenge

Get the Best Deal

Grocery stores often sell items in different quantities, package sizes, and unit prices. A unit price is the price for one unit of an item. To get the best deal, you should buy each item with the lowest unit price. Find each unit price and determine the best deal.

	1 for $0.69	6 for $2.70	12 for $4.80
	Unit price (per pound) ________	Unit price (per pound) ________	Unit price (per pound) ________

Best deal: ______________________

	1 pound for $0.75	2 pounds for $1.70	5 pounds for $4.05
	Unit price (per pound) ________	Unit price (per pound) ________	Unit price (per pound) ________

Best deal: ______________________

	6-ounce box for $1.98	12-ounce box for $3.72	16-ounce box for $5.28
FLAKO'S	Unit price (per ounce) ________	Unit price (per ounce) ________	Unit price (per ounce) ________

Best deal: ______________________

	6-pack for $1.08	12-pack for $2.64	24-pack for $4.08
FUZZO POP	Unit price (per can) ________	Unit price (per can) ________	Unit price (per can) ________

Best deal: ______________________

Name ______________________ Date __________ Class __________

LESSON 3-6

Problem Solving

Dividing Decimals by Whole Numbers

Write the correct answer.

1. Four friends had lunch together. The total bill for lunch came to $33.40, including tip. If they shared the bill equally, how much did they each pay?

____

2. There are 7.2 milligrams of iron in a dozen eggs. Because there are 12 eggs in a dozen, how many milligrams of iron are in 1 egg?

____

3. Kyle bought a sheet of lumber 8.7 feet long to build fence rails. He cut the strip into 3 equal pieces. How long is each piece?

____

4. An albatross has a wingspan greater than the length of a car—3.7 meters! Wingspan is the length from the tip of one wing to the tip of the other wing. What is the length of each albatross wing (assuming wing goes from center of body)?

____

Circle the letter of the correct answer.

5. The City Zoo feeds its three giant pandas 181.5 pounds of bamboo shoots every day. Each panda is fed the same amount of bamboo. How many pounds of bamboo does each panda eat every day?

A 6.05 pounds
B 60.5 pounds
C 61.5 pounds
D 605 pounds

6. Emma bought 22.5 yards of cloth to make curtains for two windows in her apartment. She used the same amount of cloth on each window. How much cloth did she use to make each set of curtains?

F 1.125 yards
G 10.25 yards
H 11.25 yards
J 11.52 yards

7. Aerobics classes cost $153.86 for 14 sessions. What is the fee for one session?

A $10.99
B $1.99
C about $25.00
D about $20.00

8. An entire apple pie has 36.8 grams of saturated fat. If the pie is cut into 8 slices, how many grams of saturated fat are in each slice?

F 4.1 grams
G 0.46 grams
H 4.6 grams
J 4.11 grams

Name ______________________ Date __________ Class __________

LESSON 3-6

Reading Strategies
Use a Visual Tool

You can use a hundred grid to show division with decimals.

The grid shows 0.15. →

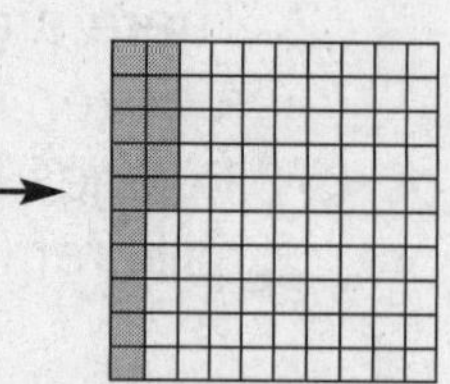

0.15 ÷ 3 means "separate 0.15 into 3 equal groups." →

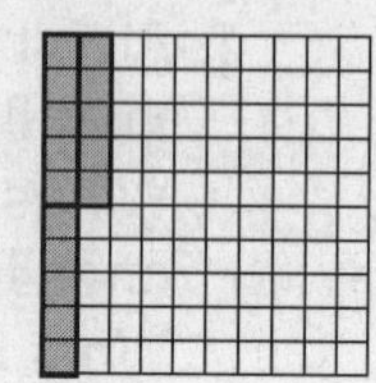

0.15 ÷ 3 makes 3 equal groups of 0.05. →

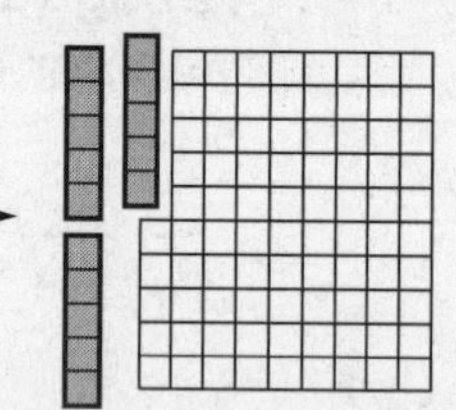

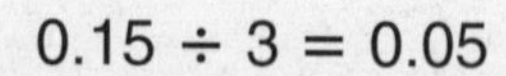

0.15 ÷ 3 = 0.05

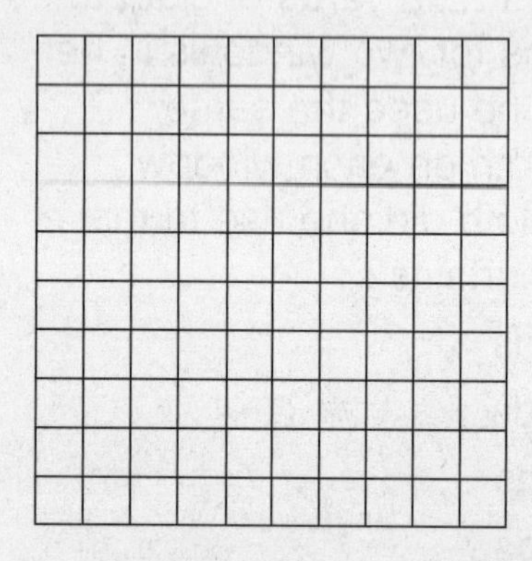

Use the grid to complete the Exercises 1–4.

1. Shade 0.60 of the grid.

2. Divide the grid into 3 equal groups.

3. Write the decimal amount in each of the 3 groups. __________

4. Write a division problem for the picture you have created.

____

Name ______________________ Date __________ Class __________

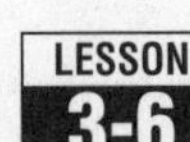

Puzzles, Twisters & Teasers

One Nation, Indivisible. . .

What state is round on both ends and raised in the middle?

We know the United States consists of fifty individual states. Often states want to make improvements to roads, state parks, or schools. They look to the federal government for monies to pay for those improvements.

For each of the improvements listed below, the government has decided which states will receive money. Divide the money for each item by the number of states receiving the money to find how much money each state gets.

ITEM	MONEY	RECIPIENT STATES	MONEY PER STATE
State Parks	$193.45 per acre	CT, CO, WY, UT, AL	
Farming	$639.24 per acre	NE, KS, MO	
Roads	$534.36 per mile	WV, GA, MS, NC, OR, WA	
Security	$94.56 per government building	MD, PA	
Schools	$143.04 per classroom	LA, OK, ME, RI, KY, AL, WI, HI	
Science/ Research	$193.20 per laboratory	FL, TX, MA, CA, AZ	
Museums	$884.79 per display	NY, NM, NH	

To answer the riddle, find the column labeled with largest amount of money and the row labeled with the smallest amount of money.

	$305.17	$296.40	$294.93	$213.08
$22.04	Maryland	North Dakota	Oklahoma	Oregon
$14.45	Ohio	Florida	Texas	Vermont
$17.88	Iowa	Minnesota	New York	Louisiana
$38.69	Pennsylvania	Arkansas	Maine	Arizona

So, what state do you think is round on both ends and raised in the middle?

____ ____ ____ ____

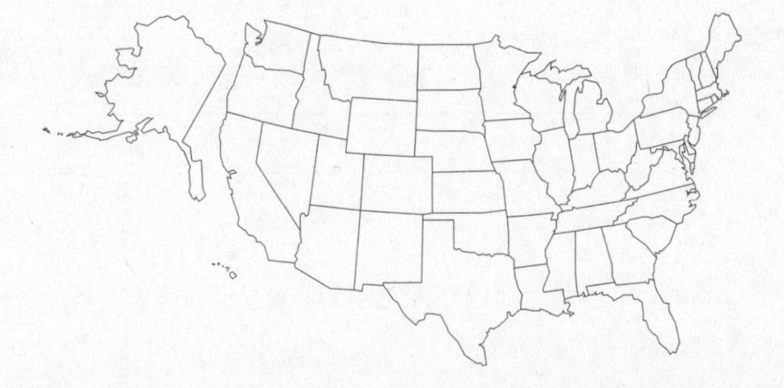

Name ______________________ Date ___________ Class ___________

LESSON 3-7

Practice A

Dividing by Decimals

Find each quotient.

1. $2.4 \div 0.4$ ________

2. $1.4 \div 0.2$ ________

3. $4.8 \div 0.6$ ________

4. $8.1 \div 0.9$ ________

5. $1.8 \div 0.3$ ________

6. $6.4 \div 0.8$ ________

7. $3.3 \div 0.3$ ________

8. $2.6 \div 1.3$ ________

9. $7.2 \div 1.2$ ________

10. $7.5 \div 1.5$ ________

11. $6.0 \div 0.5$ ________

12. $9.9 \div 1.1$ ________

Evaluate $4.8 \div x$ for each value of x.

13. $x = 0.2$ ________

14. $x = 0.4$ ________

15. $x = 0.3$ ________

16. $x = 0.6$ ________

17. $x = 0.8$ ________

18. $x = 1.2$ ________

19. Antonio spent $5.60 on cashews. They cost $1.40 per pound. How many pounds of cashews did Antonio buy?

____

20. Over several months, a scientist measured a total of 6.3 inches of snow. The average snowfall each month was 2.1 inches. How many months did the scientist measure the snow?

____

Name ______________ Date ________ Class ________

LESSON 3-7

Practice B

Dividing by Decimals

Find each quotient.

1. 9.0 ÷ 0.9 ________

2. 29.6 ÷ 3.7 ________

3. 10.81 ÷ 2.3 ________

4. 10.5 ÷ 1.5 ________

5. 15.36 ÷ 4.8 ________

6. 9.75 ÷ 1.3 ________

7. 20.4 ÷ 5.1 ________

8. 37.5 ÷ 2.5 ________

9. 9.24 ÷ 1.1 ________

10. 16.56 ÷ 6.9 ________

11. 28.9 ÷ 8.5 ________

12. 14.35 ÷ 0.7 ________

Evaluate *x* ÷ 1.2 for each value of *x*.

13. $x = 40.8$ ________

14. $x = 1.8$ ________

15. $x = 10.8$ ________

16. $x = 14.4$ ________

17. $x = 4.32$ ________

18. $x = 0.06$ ________

19. Anna is saving $6.35 a week to buy a computer game that costs $57.15. How many weeks will she have to save to buy the game?

____

20. Ben ran a 19.5-mile race last Saturday. His average speed during the race was 7.8 miles per hour. How long did it take Ben to finish the race?

____

Name ______________________ Date __________ Class __________

LESSON 3-7

Practice C

Dividing by Decimals

Find each quotient.

1. $4.75 \div 2.5$

2. $34.04 \div 4.6$

3. $10.0 \div 1.25$

4. $283.62 \div 8.7$

5. $168.75 \div 6.75$

6. $0.1092 \div 0.013$

7. $7.7293 \div 3.7$

8. $97.206 \div 5.1$

9. $0.489807 \div 0.081$

Evaluate $15.65 \div x$ for each value of x.

10. $x = 0.2$

11. $x = 0.4$

12. $x = 0.5$

13. $x = 0.8$

14. $x = 1.6$

15. $x = 2.5$

Evaluate.

16. $n \div 7.8$ for $n = 26.988$

17. $(7^2 - 32.9) \div c$ for $c = 3.5$

18. $18.67 - (0.216 \div x)$ for $x = 0.02$

19. $4.4t \div 1.6$ for $t = 16.92$

20. The sum of two decimal numbers is 3.9. Their difference is 0.9, and their product is 3.6. What are the two numbers?

21. The sum of two decimal numbers is 5.3. Their difference is 1.7, and their product is 6.3. What are the two numbers?

Name ______________________ Date __________ Class __________

LESSON 3-7

Reteach

Dividing by Decimals

You can use powers of ten to help you divide a decimal by a decimal.

To divide 0.048 by 0.12, first multiply each number by the least power of ten that makes the divisor a whole number.

$0.048 \div 0.12$

$0.12 \cdot 10^2 = 12$ Move the decimal point 2 places to the right.

$0.048 \cdot 10^2 = 4.8$ Move the decimal point 2 places to the right.

Then divide.

$4.8 \div 12$ **Step 1:** Divide as you would divide a whole number by a whole number.

$$\begin{array}{r} 0.4 \\ 12\overline{)4.8} \\ \underline{4\,8} \\ 0 \end{array}$$

Step 2: Think $48 \div 12 = 4$.

Step 3: Bring the decimal into the quotient and add a zero placeholder if necessary.

So, $0.048 \div 0.12 = 0.4$.

Find each quotient.

1. $0.7\overline{)0.42}$ **2.** $0.08\overline{)0.4}$ **3.** $0.5\overline{)0.125}$ **4.** $0.02\overline{)0.3}$

5. $0.4\overline{)0.08}$ **6.** $0.9\overline{)0.63}$ **7.** $0.008\overline{)0.4}$ **8.** $0.04\overline{)0.032}$

9. $0.3\overline{)0.06}$ **10.** $0.04\overline{)0.2}$ **11.** $0.007\overline{)4.9}$ **12.** $0.6\overline{)0.012}$

Name ______________________ Date ____________ Class ____________

LESSON 3-7

Challenge

Cutting Decimals

The strips of cloth below need to be cut into equal pieces of given lengths. Draw lines on each strip of cloth to show how many pieces will be cut.

1. Total Length: 9.8 yards   Piece Length: 1.4 yards

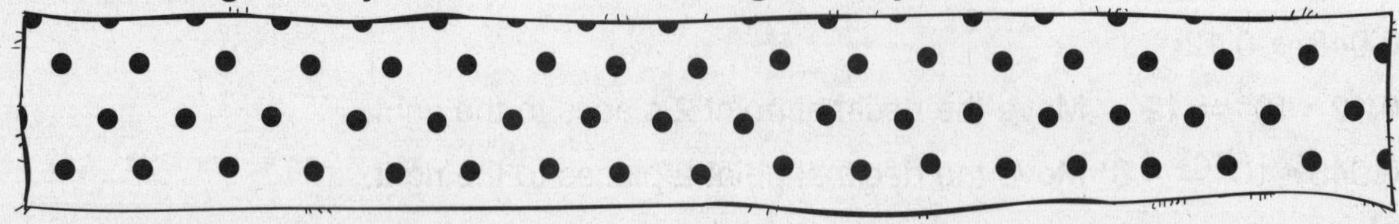

2. Total Length: 2.5 yards   Piece Length: 0.5 yards

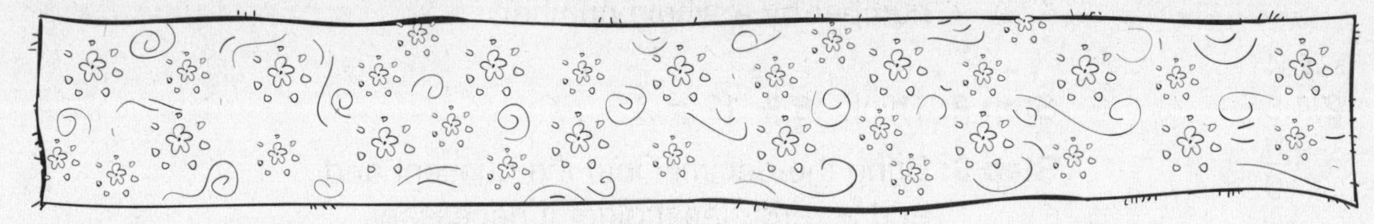

3. Total Length: 10.2 yards   Piece Length: 1.7 yards

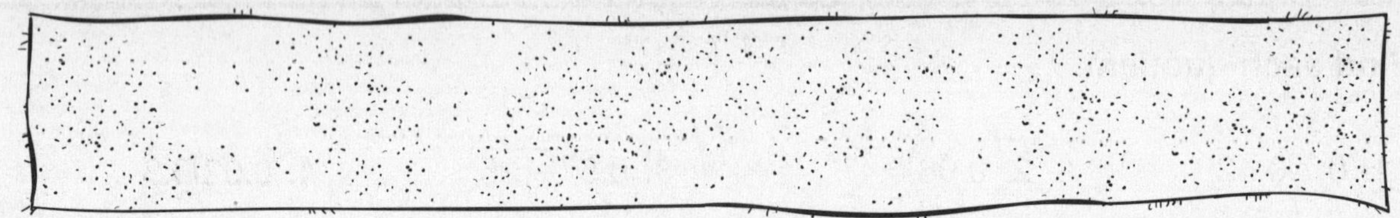

4. Total Length: 6.4 yards   Piece Length: 0.8 yards

5. Total Length: 13.6 yards   Piece Length: 3.4 yards

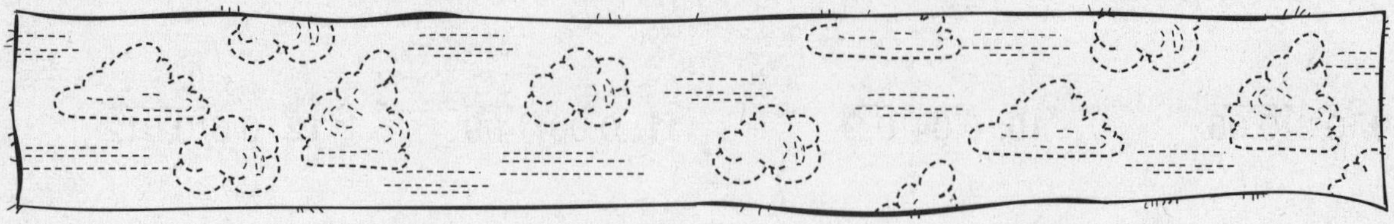

Name ______________________ Date ____________ Class ____________

LESSON **3-7**

Problem Solving

Dividing by Decimals

Write the correct answer.

1. Jamal spent $6.75 on wire to build a rabbit hutch. Wire costs $0.45 per foot. How many feet of wire did Jamal buy?

____

2. Peter drove 195.3 miles in 3.5 hours. On average, how many miles per hour did he drive?

____

3. Lisa's family drove 830.76 miles to visit her grandparents. Lisa calculated that they used 30.1 gallons of gas. How many miles per gallon did the car average?

____

4. A chef bought 84.5 pounds of ground beef. He uses 0.5 pound of ground beef for each hamburger. How many hamburgers can he make?

____

Circle the letter of the correct answer.

5. Mark earned $276.36 for working 23.5 hours last week. He earned the same amount of money for each hour that he worked. What is Mark's hourly rate of pay?
   A $1.17
   B $10.76
   C $11.76
   D $117.60

6. Alicia wants to cover a section of her wall that is 2 feet wide and 12 feet long with mirrors. Each mirror tile is 2 feet wide and 1.5 feet long. How many mirror tiles does she need to cover that section?
   F 4 tiles
   G 6 tiles
   H 8 tiles
   J 12 tiles

7. John ran the city marathon in 196.5 minutes. The marathon is 26.2 miles long. On average, how many miles per hour did John run the race?
   A 7 miles per hour
   B 6.2 miles per hour
   C 8 miles per hour
   D 8.5 miles per hour

8. Shaneeka is saving $5.75 of her allowance each week to buy a new camera that costs $51.75. How many weeks will she have to save to have enough money to buy it?
   F 9 weeks
   G 9.5 weeks
   H 8.1 weeks
   J 8 weeks

Name ______________________ Date __________ Class __________

LESSON 3-7

Reading Strategies

Make Predictions

Study the examples below. Look for patterns in the divisor and quotient.

Dividend		Divisor		Quotient
400	÷	20	=	20
400	÷	2	=	200
400	÷	0.2	=	2,000
400	÷	0.02	=	20,000

As the divisor is divided by 10, the quotient is multiplied by 10.

Use the information above to answer the Exercises 1–3.

1. Predict the divisor for the next problem in this pattern.

____

2. Predict the quotient for the next problem in this pattern.

____

3. Write the next division problem and quotient for this pattern.

____

Study the pattern created by these division problems. Use the pattern to answer Exercises 4–6.

Dividend		Divisor		Quotient
900	÷	30	=	30
900	÷	3	=	300
900	÷	0.3	=	3,000

4. Predict the next divisor in this pattern.

____

5. Predict the next quotient in this pattern.

____

6. Write the division problem and quotient that you predict would come next.

____

Name _______________ Date __________ Class __________

LESSON 3-7

Puzzles, Twisters & Teasers

Bargain Hunter

You are a professional shopper. You've been asked to use your shopping skills to find the best buy among the following cereals. You must determine which of these cereals costs the least per ounce.

Type of Cereal	Size of Box	Price per Box
Frostee O's	22.5 oz	$12.60
Super Sugar Loops	9.9 oz	$2.97
Marshmallow Bonanza	22.8 oz	$3.42
Tyrannosaurus Rings	6.6 oz	$3.30
Wheat and Rice Explosion	35.6 oz	$3.56
Giant Frosted Oatees	10.0 oz	$6.10
Ned's Enormous O's of Sugar	15.5 oz	$1.86

Type of Cereal	Price per Oz
Frostee O's	
Super Sugar Loops	
Marshmallow Bonanza	
Tyrannosaurus Rings	
Wheat and Rice Explosion	
Giant Frosted Oatees	
Ned's Enormous O's of Sugar	

Do you know what lies at the bottom of the ocean and twitches?

To find out, fill in the capital letters from the *second* best bargain cereal in order in the spaces below. Then fill in the capital letters in order from the *best* bargain cereal and you should have the answer!

A ____ ____ R V ____ U ____

____ ____ ____ C K!

Name ________________________ Date ____________ Class ____________

LESSON 3-8

Practice A

Interpret the Quotient

Circle the letter of the correct answer.

1. Hamburger rolls come in packs of 8. How many packs should you buy to have 60 rolls?

A 8

B 6

C 5

D 7

2. Each pack of hamburger rolls costs $1.50. How many packs can you buy with $8.00?

F 6

G 5

H 4

J 8

3. How many 0.6-pound hamburgers can you make with 7.8 pounds of ground beef?

A 13

B 14

C 10

D 16

4. You spend a total of $5.10 for 3 pounds of ground beef. How much does the ground beef cost per pound?

F $0.70

G $0.17

H $15.30

J $1.70

Write the correct answer.

5. Four friends equally shared the cost of buying supplies for the class picnic. The supplies cost a total of $12.40. How much did they each pay?

____

6. In all, 20 people are going to the picnic. Each van seats 6 people. How many vans are needed to take everyone to the picnic?

____

7. Plastic forks come in packs of 6. If you need 40 forks for the picnic, how many packs should you buy?

____

8. You spent a total of $9.60 on paper plates for the picnic. Each pack costs $1.20. How many packs of paper plates did you buy?

____

Name ______________________ Date __________ Class __________

Practice B
Interpret the Quotient

Circle the letter of the correct answer.

1. You spent a total of $6.75 for 15 yards of ribbon. How much did the ribbon cost per yard?

A $0.50
B $0.45
C $1.35
D $1.45

2. Buttons come in packs of 12. How many packs should you buy if you need 100 buttons?

F 10
G 8
H 9
J 12

3. Your sewing cabinet has compartments that hold 8 spools of thread each. You have 50 spools of thread. How many compartments can you fill?

A 6
B 7
C 5
D 8

4. You spent a total of $35.75 for velvet cloth. Each yard of the velvet costs $3.25. How many yards did you buy?

F 10
G 10.5
H 11
J 11.5

Write the correct answer.

5. You used a total of 67.5 yards of cotton material to make costumes for the play. Each costume used 11.25 yards of cloth. How many costumes did you make?

____

6. You are saving $17.00 each week to buy a new sewing machine that costs $175.50. How many weeks will you have to save to have enough money to buy the sewing machine?

____

7. Sequins come in packs of 75. You use 12 sequins on each costume. If you have one pack of sequins, how many costumes can you make?

____

8. You pay $26.28 for a subscription to *Sewing Magazine.* You get an issue every month for a year. How much does each issue cost?

____

Name ____________________ Date __________ Class __________

LESSON 3-8

Practice C

Interpret the Quotient

Write the correct answer.

1. You live in Detroit, Michigan. You and your parents will be driving to Cincinnati, Ohio, to visit your grandparents. The trip is a total of 264.1 miles. Your family car averages 27.8 miles per gallon. How many gallons of gas will you use on the trip?

____

2. You plan on taking 100 photographs during your trip. You want to choose one kind of film to buy, either a 24-photo roll, or an 18-photo roll. How many packs of each would you need to buy?

____

3. The gas tank of your parents' car holds 12.9 gallons of gas. Because their car averages 27.8 miles per gallon, how many times will they have to fill up the gas tank to drive to Cincinnati and back?

____

4. The first time you stop at a gas station during your trip, your parents spend $13.80 for 11.5 gallons of gas. How much does the gas cost per gallon?

____

5. You stop for lunch during the trip and spend a total of $11.38. Your cheeseburger costs $3.25. Both of your parents have a slice of pizza that each cost $2.19. You all get the same drink. How much did each drink cost?

____

6. During the trip your parents drive 211.05 miles on different highways. You calculate that you spent 3.5 hours driving on highways. What was the average speed your parents drove during that part of the trip?

____

7. You saved for 16 weeks before your trip to buy a present for your grandparents that cost $67.36. How much did you save each week?

____

8. Your parents budgeted $20 a month for long-distance calls. It costs $0.37 per minute for long distance calls. How many minutes can you talk to your grandparents each month?

____

Name ______________________________ Date ___________ Class ___________

Reteach

Interpret the Quotient

There are three ways the decimal part of a quotient can be interpreted when you solve a problem.

If the question asks for an exact number, use the entire quotient.
If the question asks how many whole groups are needed to put the dividend into a group, round the quotient up to the next whole number.
If the question asks how many whole groups can be made when you divide, drop the decimal part of the quotient.

To interpret the quotient, decide what the question is asking.

In the school library, there are tables that seat 4 students each. If there are 30 students in a class, how many tables are needed to seat all of the students?

To solve, divide 30 by 4.

$30 \div 4 = 7.5$

The question is asking how many tables (whole groups) are needed to put all of the students in the class (dividend) into a group.

So, round 7.5 up to the next whole number.

8 tables are needed to seat all of the students.

Interpret the quotient to solve each problem.

1. A recipe that serves 6 requires 9 cups of milk. How much milk is needed for each serving?

____

2. A storage case holds 24 model cars. Marla has 84 model cars. How many storage cases does she need to store all of her cars?

____

3. Kenny has $4.25 to spend at the school carnival. If game tickets are $0.50 each, how many games can Kenny play?

____

Name ______________________________ Date ___________ Class ___________

LESSON 3-8

Challenge

Plan a Party!

You are in charge of buying supplies for the class party. There are 30 students in your class. Use the party supply store advertisement below to plan what to buy. After you pay for all the items, the total cost will be divided evenly among all the students.

Shopping List

Item	Number of Items You Want Per Person	Number of Packs to Buy	Number of Left Over Items	Total Price of Items
Invitations	1			
Paper plates	1			
Plastic cups	2			
Paper napkins	2			
Plastic forks	1			
			Grand Total Price:	
			Cost Per Student:	

Name ______________________ Date __________ Class __________

LESSON **3-8**

Problem Solving

Interpret the Quotient

Write the correct answer.

1. Five friends split a pizza that costs \$16.75. If they shared the bill equally, how much did they each pay?

____

2. There are 45 choir members going to the recital. Each van can carry 8 people. How many vans are needed?

____

3. Tara bought 150 beads. She needs 27 beads to make each necklace. How many necklaces can she make?

____

4. Cat food costs \$2.85 for five cans. Ben only wants to buy one can. How much will it cost?

____

Circle the letter of the correct answer.

5. Tennis balls come in cans of 3. The coach needs 50 tennis balls for practice. How many cans should he order?

A 16 cans
B 17 cans
C 18 cans
D 20 cans

6. The rainfall for three months was 4.6 inches, 3.5 inches, and 4.2 inches. What was the average monthly rainfall during that time?

F 41 inches
G 12.3 inches
H 4.3 inches
J 4.1 inches

7. Tom has \$15.86 to buy marbles that cost \$1.25 each. He wants to know how many marbles he can buy. What should he do after he divides?

A Drop the decimal part of the quotient when he divides.
B Drop the decimal part of the dividend when he divides.
C Round the quotient up to the next highest whole number to divide.
D Use the entire quotient of his division as the answer.

8. Mei needs 135 hot dog rolls for the class picnic. The rolls come in packs of 10. She wants to know how many packs to buy. What should she do after she divides?

F Drop the decimal part of the quotient when she divides.
G Drop the decimal part of the dividend when she divides.
H Round the quotient up to the next highest whole number.
J Use the entire quotient of her division as the answer.

Name ______________________________ Date ____________ Class ____________

LESSON 3-8

Reading Strategies

Use Context

How the decimal portion of the quotient in a division problem is used depends upon the situation.

Situation 1 74 students are going on a field trip in cars. Each car can carry 5 students. How many cars are needed?

Divide 74 by 5. ⟶ $74 \div 5 = 14.8$ cars

Reasoning 14 cars will not be enough for all students. You need 15 cars. The quotient 14.8 needs to be **rounded up** to 15 in this situation.

Situation 2 How many 8 oz servings are in a 44 oz can of juice?

Divide 44 by 8. ⟶ $44 \div 8 = 5.5$ servings

Reasoning There are 5 full 8 oz servings in the can. The 0.5 serving is not 8 ounces. The quotient 5.5 is **rounded down** to 5 in this situation.

Situation 3 4 boys mowed a lawn for $35. How much money should each boy receive to share the money equally?

Divide $35 by 4. ⟶ $\$35 \div 4 = \$8.75$

Reasoning The exact quotient of $8.75 states what each boy should receive. The **exact quotient** of $8.75 makes sense.

Tell whether you would round the quotient up, round the quotient down, or leave the exact quotient for each. Write to explain your choice.

1. You need 8 inches of ribbon to make a bow. How many bows can you make with 50 inches of ribbon? $50 \div 8 = 6.25$

____

2. Each lunch table seats 10 children. There are 155 children in the cafeteria for each lunch period. How many tables are needed? $155 \div 10 = 15.5$

____

Name ______________________________ Date ____________ Class ____________

LESSON **3-8**

Puzzles, Twisters & Teasers

To Dine or Not to Dine?

John loves to have his friends over for dinner. He is concerned that he will not have enough food.

He knows that he has 8 cups of soup, 14 pounds of lasagna, 2.85 pounds of salad, and 7.3 gallons of ice cream. John knows that, on average, each guest eats about 0.75 cups of soup, 1.2 pounds of lasagna, 0.3 pounds of salad, and 0.6 pint of ice cream. Use these numbers to help John decide how many people he can invite for dinner.

Soup ____________

Salad ____________

Lasagna ____________

Ice Cream ____________

7	T
8	O
9	U
10	M
11	L
12	N

Number of people John should invite: ____________

Do you know what John's gossiping friend had during dinner?

Find the letter that is paired with the number of people that John should invite and add it to the letter bank. Also, add one to the number of people John can invite, find the matching letter in the table above and add that letter to the letter bank as well.

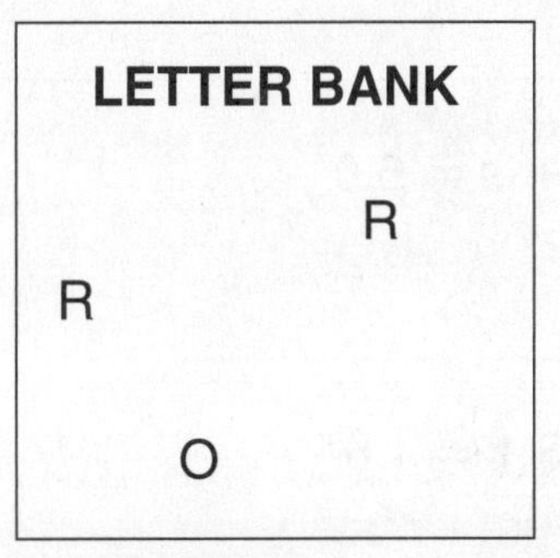

Rearrange the letters in the letter bank to find the answer to the riddle.

He had a great sense of

____ ____ ____ ____ ____.

Name ______________________ Date ____________ Class ____________

LESSON 3-9

Practice A

Solving Decimal Equations

Circle the letter of the correct answer.

1. If $7.2 + x = 9.7$, what is the value of x?

A 16.9

B 2.5

C 2.9

D 16.5

2. If $n \div 4 = 0.4$, what is the value of n?

F 4.4

G 0.1

H 1.6

J 10

Solve each equation. Check your answer.

3. $a - 0.4 = 1.3$

____

4. $2n = 1.8$

____

5. $0.8 + x = 1.3$

____

6. $p \div 4 = 0.7$

____

7. $w - 2.3 = 1.2$

____

8. $0.5q = 0.25$

____

9. $5.7 + s = 6.0$

____

10. $b \div 3 = 0.6$

____

11. $t - 3.1 = 1.6$

____

12. $3y = 1.5$

____

13. The length of a photograph is 2.1 inches, and the width is 3 inches. Solve the equation $a \div 2.1 = 3$ to find the area of the photograph.

____

14. It costs \$0.90 to enlarge each photograph. If you want to enlarge 5 photos, how much will it cost in all?

____

Name ________________ Date ________ Class ________

LESSON 3-9

Practice B

Solving Decimal Equations

Solve each equation. Check your answer.

1. $a - 2.7 = 4.8$

______

2. $b \div 7 = 1.9$

______

3. $w - 6.5 = 3.8$

______

4. $p \div 0.4 = 1.7$

______

5. $4.5 + x = 8$

______

6. $b \div 3 = 2.5$

______

7. $7.8 + s = 15.2$

______

8. $1.63q = 9.78$

______

9. $0.05 + x = 2.06$

______

10. $1.7n = 2.38$

______

11. $t - 6.08 = 12.59$

______

12. $9q = 16.2$

______

13. $w - 8.9 = 10.3$

______

14. $1.4n = 3.22$

______

15. $t - 12.7 = 0.8$

______

16. $3.8 + a = 6.5$

______

17. The distance around a square photograph is 12.8 centimeters. What is the length of each side of the photograph?

______

18. You buy two rolls of film for $3.75 each. You pay with a $10 bill. How much change should you get back?

______

Name ______________________ Date __________ Class __________

LESSON 3-9

Practice C

Solving Decimal Equations

Solve each equation. Check your answer.

1. $a - 0.089 = 12.5$

____

2. $b \div 2.8 = 4.7$

____

3. $w - 12.76 = 15.8$

____

4. $p \div 0.07 = 4.76$

____

5. $1.7806 + x = 2.009$

____

6. $b \div 6.5 = 9.7$

____

7. $8.3044 + s = 16.01$

____

8. $0.009q = 0.765$

____

9. $23.764 + x = 30.5$

____

10. $1.7n + 3.8 = 7.71$

____

11. $t - (8^2 + 0.36) = 0.5$

____

12. $9q + 0.2q = 27.6$

____

13. $8.4w - 0.67 = 15.29$

____

14. $1.4n + 4^3 = 85$

____

15. Andy bought 15 pounds of apples for \$0.69 a pound, and 7.2 pounds of grapes for \$3.65 a pound. He paid for the entire purchase with a \$50 bill. How much change did he get back? He wants to make 25 pounds of fruit salad for the party. How much more fruit does he need?

____

____

16. Brenda's garden is 6.5 feet long and 8.76 feet wide. She uses half of the garden for vegetables. Then she equally divides the rest of the garden between flowers and herbs. How large is her vegetable garden? How much land does she use for growing flowers?

____

____

Name ______________________ Date __________ Class __________

LESSON 3-9

Reteach

Solving Decimal Equations

You can write related equations for addition and subtraction equations.

$7.4 + 6.2 = 13.6$ $\quad$ $13.6 - 6.2 = 7.4$

Use related equations to solve each of the following.

A. $x + 4.5 = 7.9$

Think: $7.9 - 4.5 = x$

$x = 3.4$

Check $x + 4.5 = 7.9$

$3.4 + 4.5 \stackrel{?}{=} 7.9$ substitute

$7.9 = 7.9$

B. $x - 0.08 = 6.2$

Think: $6.2 + 0.08 = x$

$x = 6.28$

Check $x - 0.08 = 6.2$

$6.28 - 0.08 \stackrel{?}{=} 6.2$ substitute

$6.2 = 6.2$

Use related facts to solve each equation. Then check each answer.

1. $x + 8.7 = 12.9$ $\quad$ **2.** $x + 8.4 = 16.6$ $\quad$ **3.** $x - 2.65 = 7.8$ $\quad$ **4.** $x - 0.8 = 2.3$

________ ________ ________ ________

You can write related equations for multiplication and division equations.

$3.2 \bullet 2.4 = 7.68$ $\quad$ $7.68 \div 2.4 = 3.2$

Use related equations to solve each of the following.

C. $3x = 1.5$

Think: $1.5 \div 3 = x$

$x = 0.5$

Check: $3x = 1.5$

$3 \bullet 0.5 \stackrel{?}{=} 1.5$ substitute

$1.5 = 1.5$

D. $x \div 6 = 1.2$

Think: $1.2 \bullet 6 = x$

$x = 7.2$

Check: $x \div 6 = 1.2$

$7.2 \div 6 \stackrel{?}{=} 1.2$ substitute

$1.2 = 1.2$

Use related facts to solve each equation. Then check each answer.

5. $x \div 3 = 6.3$ $\quad$ **6.** $x \div 0.2 = 3.4$ $\quad$ **7.** $7x = 4.2$ $\quad$ **8.** $5x = 4.5$

________ ________ ________ ________

Name ______________________ Date ____________ Class ____________

LESSON 3-9

Challenge

Playing Weight

In professional sports, each ball has a maximum, or greatest, weight allowed in play. The lightest official weight for a table tennis ball is only 0.009 ounces. Use the equations below to find the maximum weights, in ounces, of some other sports' balls.

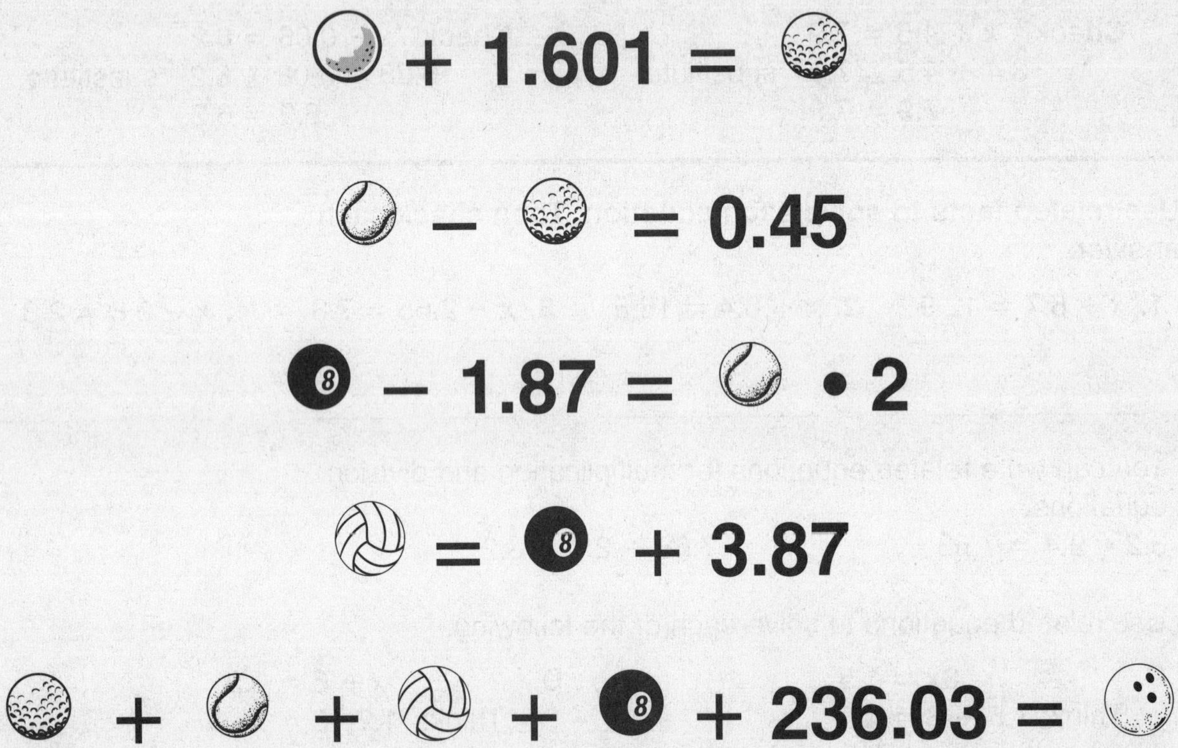

+ 1.601 =

− = 0.45

− 1.87 = • 2

= + 3.87

+ + + + 236.03 =

Golf Ball

Weight: ____________

Tennis Ball

Weight: ____________

Billiard Ball

Weight: ____________

Volleyball

Weight: ____________

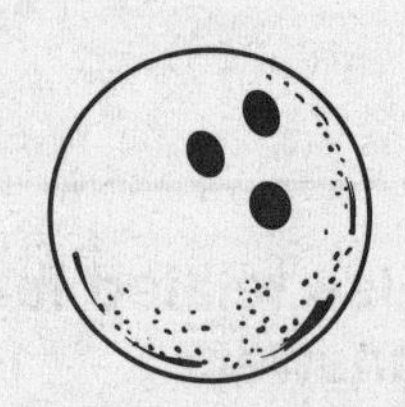

Bowling Ball

Weight: ____________

Name ______________________ Date ____________ Class ____________

LESSON 3-9

Problem Solving

Solving Decimal Equations

Write the correct answer.

1. Bee hummingbirds weigh only 0.0056 ounces. They have to eat half their body weight every day to survive. How much food does a bee hummingbird have to eat each day?

____

2. The desert locust, a type of grasshopper, can jump 10 times the length of its body. The locust is 1.956 inches long. How far can it jump in one leap?

____

3. In 1900, there were about 1.49 million people living in California. In 2000, the population was 33.872 million. How much did the population grow between 1900 and 2000?

____

4. Juanita has $567.89 in her checking account. After she deposited her paycheck and paid her rent of $450.00, she had $513.82 left in the account. How much was her paycheck?

____

Circle the letter of the correct answer.

5. The average body temperature for people is 98.6°F. The average body temperature for most dogs is 3.4°F higher than for people. The average body temperature for cats is 0.5°F lower than for dogs. What is the normal body temperature for dogs and cats?

   A dogs: 101.5°F; cats 102°F

   B dogs: 102°F; cats 101.5°F

   C dogs: 102.5°F; cats 103°F

   D dogs: 102.5°F; cats 102.5°F

6. Seattle, Washington, is famous for its rainy climate. Winter is the rainiest season there. From November through December the city gets an average of 5.85 inches of rain each month. Seattle usually gets 6 inches of rain in December. What is the city's average rainfall in November?

   F 6 inches

   G 5.925 inches

   H 5.8 inches

   J 5.7 inches

7. The equation to convert from Celsius to Kelvin degrees is $K = 273.16 + C$. If it is 303.66°K outside, what is the temperature in Celsius degrees?

   A 576.82°C

   B 30.5°C

   C 305°C

   D 257.68°C

8. The distance around a square mirror is 6.8 feet. Which of the following equations finds the length of each side of the mirror?

   F $6.8 - x = 4$

   G $x \div 4 = 6.8$

   H $4x = 6.8$

   J $6.8 + 4 = x$

Name ______________________ Date __________ Class __________

LESSON 3-9

Reading Strategies

Use a Flowchart

This flowchart can help you work with decimal equations.

1. How do I read this equation?

↓

2. What operation is shown in this equation?

↓

3. What is the inverse operation for this equation?

Adding or Subtracting Decimals

$x + 9.7 = 15.4$ ← **1.** Read "*x* plus 9.7 equals 15.4."

$x + \mathbf{9.7} = 15.4$ ← **2.** Addition is shown.

$x + 9.7 - \mathbf{9.7} = 15.4 - \mathbf{9.7}$ ← **3.** The inverse operation is subtraction.

Use this equation to complete Exercises 1– 2: $n - 4.5 = 6.3$.

1. Write in words how you read the equation.

____

2. What operation is shown in this equation? What is the inverse of that operation?

____

Use this equation to complete Exercises 3–4: $w + 9 = 4.8$.

3. Write in words how you read this equation.

____

4. What operation is shown in this equation? What is the inverse of that operation?

____

Name ______________________________ Date ____________ Class ____________

LESSON **3-9**

Puzzles, Twisters & Teasers

Super Cross

The Super Cross is the empty grid below. Fill in the grid with the letters from the equations below to answer the riddle.

First, solve all of the equations. Then enter the letter with the highest value in the top left corner. The next highest goes in the space to its right and then continue filling in the numbers from highest to lowest, left to right and top to bottom.

Read the answer to the riddle by reading down the rightmost column, then down the center column, then down the leftmost column.

$4.9b = 0.637$

$l - 54 = 45.1$

$\frac{h}{5} = 6.5$

$i + 9.3 = 15.6$

$4s = 2.8$

$65.7 - t = 24.9$

$a + 4.5 = 21.7$

$\frac{e}{5.6} = 3.2$

$34.2s = 3.42$

b	0.13
l	99.1
h	32.5
i	6.3
s	0.7
t	40.8
a	17.2
e	17.92
s	0.1

Do you know why the waiter was good at multiplication?

Because he knew

____ ____ ____

____ ____ ____ ____ ____ ____!

Name ______________________ Date ____________ Class ____________

CHAPTER 3

Teacher Tools

Set of Cards 0-9

0	1	2
3	4	5
6	7	8
9		

LESSON 3-1

Practice A

Representing, Comparing, and Ordering Decimals

Write the value of the underlined digit in each number.

1. 1.6 — 6 tenths
2. 7.62 — 7 ones
3. 3.69 — 9 hundredths
4. 20.4 — 2 tens
5. 5.136. — 1 tenth
6. 5.08 — 8 hundredths

Write each decimal in standard form, expanded form, and words.

7. 1.8 — 1 + 0.8; one and eight tenths
8. 3 + 0.6 + 0.02 — 3.62; three and sixty-two hundredths
9. one and fifty-two hundredths — 1.52; 1 + 0.5 + 0.02

Circle the letter of the correct answer.

10. Which of the following sets is written in order from greatest to least?
    A 1.7, 1.07, 17
    (B) 5.2, 2.5, 0.52
    C 1.07, 17, 1.7
    D 2.5, 0.52, 5.2

11. Which of the following sets is written in order from least to greatest?
    F 0.85, 8.5, 5.8
    G 4.3, 3.4, 0.43
    H 5.8, 0.85, 8.5
    (J) 0.43, 3.4, 4.3

12. Reno, Nevada, gets an average of only five-tenths inch of rain in June, and only three-tenths inch of rain in July. Which month in Reno has less rain?

    July

13. Honolulu, Hawaii, gets an average of three and eight tenths inches of rain in December, and three and six tenths inches of rain in January. Which month in Honolulu has more rain?

    December

LESSON 3-1

Practice B

Representing, Comparing, and Ordering Decimals

Write each decimal in standard form, expanded form, and words.

1. 2.07 — 2 + 0.07; two and seven hundredths
2. 5 + 0.007 — 5.007; five and seven thousandths
3. four and six tenths — 4.6; 4 + 0.6
4. sixteen and five tenths — 16.5; 10 + 6 + 0.5
5. 9 + 0.6 + 0.08 — 9.68; nine and sixty-eight hundredths
6. 1.037 — 1 + 0.03 + 0.007; one and thirty-seven thousandths
7. 2 + 0.1 + 0.003 — 2.103; two and one hundred three thousandths
8. eighteen hundredths — 0.18; 0.1 + 0.08
9. 6.11 — 6 + 0.1 + 0.01; six and eleven hundredths

Order the decimals from least to greatest.

10. 3.578, 3.758, 3.875
    3.578; 3.758; 3.875
11. 0.0943, 0.9403, 0.9043
    0.0943; 0.9043; 0.9403
12. 12.97, 12.957, 12.75
    12.75; 12.957; 12.97
13. 1.09, 1.901, 1.9, 1.19
    1.09; 1.19, 1.9; 1.901

14. Your seventh and eighth ribs are two of the longest bones in your body. The average seventh rib is nine and forty-five hundredths inches long, and the average eighth rib is 9.06 inches long. Which bone is longer?

    the seventh rib

15. The average female human heart weighs nine and three tenths ounces, while the average male heart weighs eleven and one tenth ounces. Which human heart weighs less, the male or the female?

    the female heart

16. The state has $42.3 million for a new theater. The theater that an architect designed would cost $42.25 million. Can the theater be built for the amount the state can pay?

    yes

17. Lyn traveled 79.47 miles on Saturday, 54.28 miles on Sunday, 65.5 miles on Monday, and 98.43 miles on Tuesday. Which day did she travel the greatest number of miles?

    Tuesday

LESSON 3-1

Practice C

Representing, Comparing, and Ordering Decimals

Compare. Write <, >, or =.

1. 10.569 [>] 10 + 0.05 + 0.006 + 0.0009
2. seventy-five hundredths [<] 7.50
3. twenty thousandths [<] twelve hundredths
4. 98.30675 [>] 90 + 8 + 0.03 + 0.007

Order from least to greatest.

5. 12.8962, 12.9682, 12.8692 — 12.8692; 12.8962; 12.9682
6. 8.098, 7.098, 8.079, 7.089 — 7.089; 7.098; 8.079; 8.098
7. 65.21, 6.521, 6.0521, 65.12 — 6.0521; 6.521; 65.12; 65.21
8. 0.304, 0.30, 0.403, 0.43, 0.34 — 0.30; 0.304; 0.34; 0.403; 0.43

Order from greatest to least.

9. 9.653, 90.563, 90.6053 — 90.6053; 90.563; 9.653
10. 11.717, 11.771, 11.117, 11.171 — 11.771; 11.717; 11.171; 11.117
11. 8.0359, 8.3509, 8.359, 8.5 — 8.5; 8.359; 8.3509; 8.0359
12. 2.35, 2.05, 2.03, 2.30, 2.53, 2.5 — 2.53; 2.5; 2.35; 2.30; 2.05; 2.03

13. From 1984 to 1996 American Carl Lewis won every men's gold medal for the long jump at the Summer Olympic Games. In 1984, he jumped 8.54 meters. In 1988, he jumped 8.72 meters. In 1992, he jumped 8.67 meters, and in 1996, he jumped 8.5 meters. Write Lewis' long jumps in order from least to greatest distance. In which year did he jump the farthest?

    8.5, 8.54, 8.67, 8.72; in 1988

14. The world records for the four highest pole vaults are 6.01 meters, 6.03 meters, 6.0 meters, and 6.14 meters. Write the records in order from greatest to least height. What is the world record for the highest pole vault?

    6.14, 6.03, 6.01, 6.0; 6.14 meters

LESSON 3-1

Reteach

Representing, Comparing, and Ordering Decimals

You can use place value to write decimals in standard form, expanded form, and word form.

To write 2.14 in expanded form, write the decimal as an addition expression using the place value of each digit.

Ones	Tenths	Hundredths	Thousandths	Ten Thousandths
2.	1	4		

2.14 can be written as 2 + 0.1 + 0.04.

When you write a decimal in word form, the number before the decimal point tells you how many wholes there are. The decimal point stands for the word "and."

Notice that the place value names to the right of the decimal begin with tenths, hundredths, and then thousandths. The "ths" ending indicates a decimal.

2.14 can also be written as *two and fourteen hundredths.*

1. How would you read a number with 4 decimal places?

   The decimal should end with ten thousandths

Write each decimal in standard form, expanded form, and word form.

2.

Ones	Tenths	Hundredths	Thousandths	Ten Thousandths
5.	6	9	8	

5 + 0.6 + 0.09 + 0.008; five and six hundred ninety-eight thousandths

3.

Ones	Tenths	Hundredths	Thousandths	Ten Thousandths
0.	0	9	4	

0 + 0.09 + 0.004; ninety-four thousandths

4. 7 + 0.8

   7.8; seven and eight tenths

5. twelve-hundredths

   0 + 0.1 + 0.02; 0.12

LESSON 3-1
Reteach
Representing, Comparing, and Ordering Decimals (cont.)

You can use place value to compare decimals.
Use < or > to compare the decimals.

Ones	Tenths	Hundredths	Thousandths	Ten Thousandths
3.	7	6	8	
3.	7	5	4	

0.06 > 0.05, so 3.768 > 3.754.

Compare. Write >, <, or =.

6.

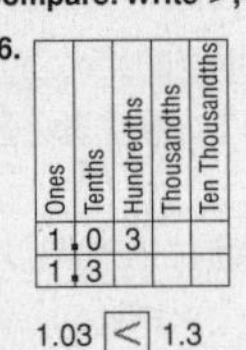

Ones	Tenths	Hundredths	Thousandths	Ten Thousandths
1.	0	3		
1.	3			

1.03 [<] 1.3

7.

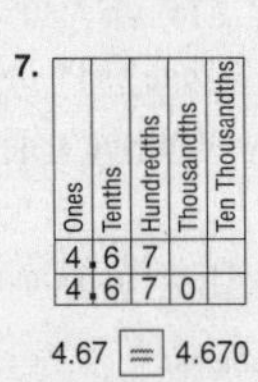

Ones	Tenths	Hundredths	Thousandths	Ten Thousandths
4.	6	7		
4.	6	7	0	

4.67 [=] 4.670

8.

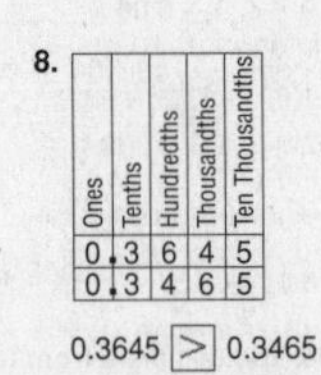

Ones	Tenths	Hundredths	Thousandths	Ten Thousandths
0.	3	6	4	5
0.	3	4	6	5

0.3645 [>] 0.3465

9. 8.53 [>] 8.053
10. 2.253 [>] 2.1345
11. 0.87 [>] 0.08703

You can use place value to order decimals.

To order 9.76, 8.59, and 9.24, from least to greatest, first compare the numbers in pairs.

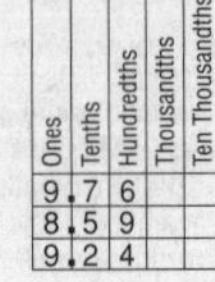

Ones	Tenths	Hundredths	Thousandths	Ten Thousandths
9.	7	6		
8.	5	9		
9.	2	4		

9.76 > 8.59, 8.59 < 9.24, 9.76 > 9.24.

So the numbers from least to greatest are 8.59, 9.24, 9.76.

Write the decimals in order from least to greatest.

12. 0.54, 0.43, 0.52 — **0.43, 0.52, 0.54**
13. 3.43, 3.34, 3.4 — **3.34, 3.4, 3.43**
14. 8.9, 9.8, 9.5 — **8.9, 9.5, 9.8**
15. 0.83, 0.8, 0.083 — **0.083, 0.8, 0.83**
16. 1.1, 0.01,1.01 — **0.01, 1.01, 1.1**
17. 6.5, 6.0, 0.6 — **0.6, 6.0, 6.5**

LESSON 3-1
Challenge
Place Your Values

Complete the tables below to show different numbers that can be written with the same digits. Do not use the same digit more than once for each place value. Possible answers are given.

1. Use the digits 1, 3, 5, 7, and 9 to write four 5-digit numbers of increasing value.

Hundreds	Tens	Ones	Tenths	Hundredths	Thousandths	Ten-Thousandths
		1	3	5	7	9
		3	1	9	5	7
		5	7	3	9	1
		7	9	1	3	5

2. Use the digits 0, 2, 4, 6, 7, and 8 to write four 6-digit numbers of decreasing value.

Hundreds	Tens	Ones	Tenths	Hundredths	Thousandths	Ten-Thousandths
	8	7	6	4	0	2
	6	8	0	2	7	4
	4	0	8	7	2	6
	2	6	4	0	8	7

3. Use the digits 0, 1, 2, 3, 4, 5, and 6 to write four 7-digit numbers of increasing value.

Hundreds	Tens	Ones	Tenths	Hundredths	Thousandths	Ten-Thousandths
1	0	2	3	4	5	6
2	1	0	4	3	6	5
3	5	6	2	0	1	4
4	3	5	6	2	0	1

LESSON 3-1
Problem Solving
Representing, Comparing, and Ordering Decimals

Use the table to answer the questions.

Largest Marine Mammals

Mammal	Length (ft)	Weight (T)
Blue whale	110.0	127.95
Fin whale	82.0	44.29
Gray whale	46.0	32.18
Humpback whale	49.2	26.08
Right whale	57.4	39.37
Sperm whale	59.0	35.43

1. What is the heaviest marine mammal on Earth?
the blue whale

2. Which mammal in the table has the shortest length?
a gray whale

3. Which mammal in the table is longer than a humpback whale, but shorter than a sperm whale?
a right whale

Circle the letter of the correct answer.

4. Which mammal measures forty-nine and two tenths feet long?
   - A blue whale
   - B gray whale
   - C sperm whale
   - (D) humpback whale

5. Which mammal weighs thirty-five and forty-three hundredths tons?
   - F right whale
   - (G) sperm whale
   - H gray whale
   - J fin whale

6. Which of the following lists shows mammals in order from the least weight to the greatest weight?
   - A sperm whale, right whale, fin whale, gray whale
   - B fin whale, sperm whale, gray whale, blue whale
   - C fin whale, right whale, sperm whale, gray whale
   - (D) gray whale, sperm whale, right whale, fin whale

7. Which of the following lists shows mammals in order from the greatest length to the least length?
   - (F) sperm whale, right whale, humpback whale, gray whale
   - G gray whale, humpback whale, right whale, sperm whale
   - H right whale, sperm whale, gray whale, humpback whale
   - J humpback whale, gray whale, sperm whale, right whale

LESSON 3-1
Reading Strategies
Connect Symbols and Words

You can read and write decimals in three ways. A place value chart can help you read decimals.
When you read or say a decimal, say "and" when you come to the decimal point.

Ones	Tenths	Hundredths	Read:
2.	5		2 **and** 5 tenths
0.	1	7	17 hundredths
8.	0	6	8 **and** 6 hundredths

Use this chart to help you write decimals in standard form and in expanded form.

Words and Symbols	Standard Form	Expanded Form
2 and 5 tenths	2.5	2 + 0.5
17 hundredths	0.17	0.1 + 0.07
8 and 6 hundredths	8.06	8 + 0.06

Write each number in words and symbols, standard form, or expanded form.

1. Write 2.17 with words and symbols. **2 and 17 hundredths**
2. Write 2.17 in expanded form. **2 + 0.1 + 0.07**
3. Write 3 and 6 hundredths in standard form. **3.06**
4. Write 3 and 6 hundredths in expanded form. **3 + 0.06**
5. Write 1.5 with words and symbols. **1 and 5 tenths**
6. Write 1.5 in expanded form. **1 + 0.5**

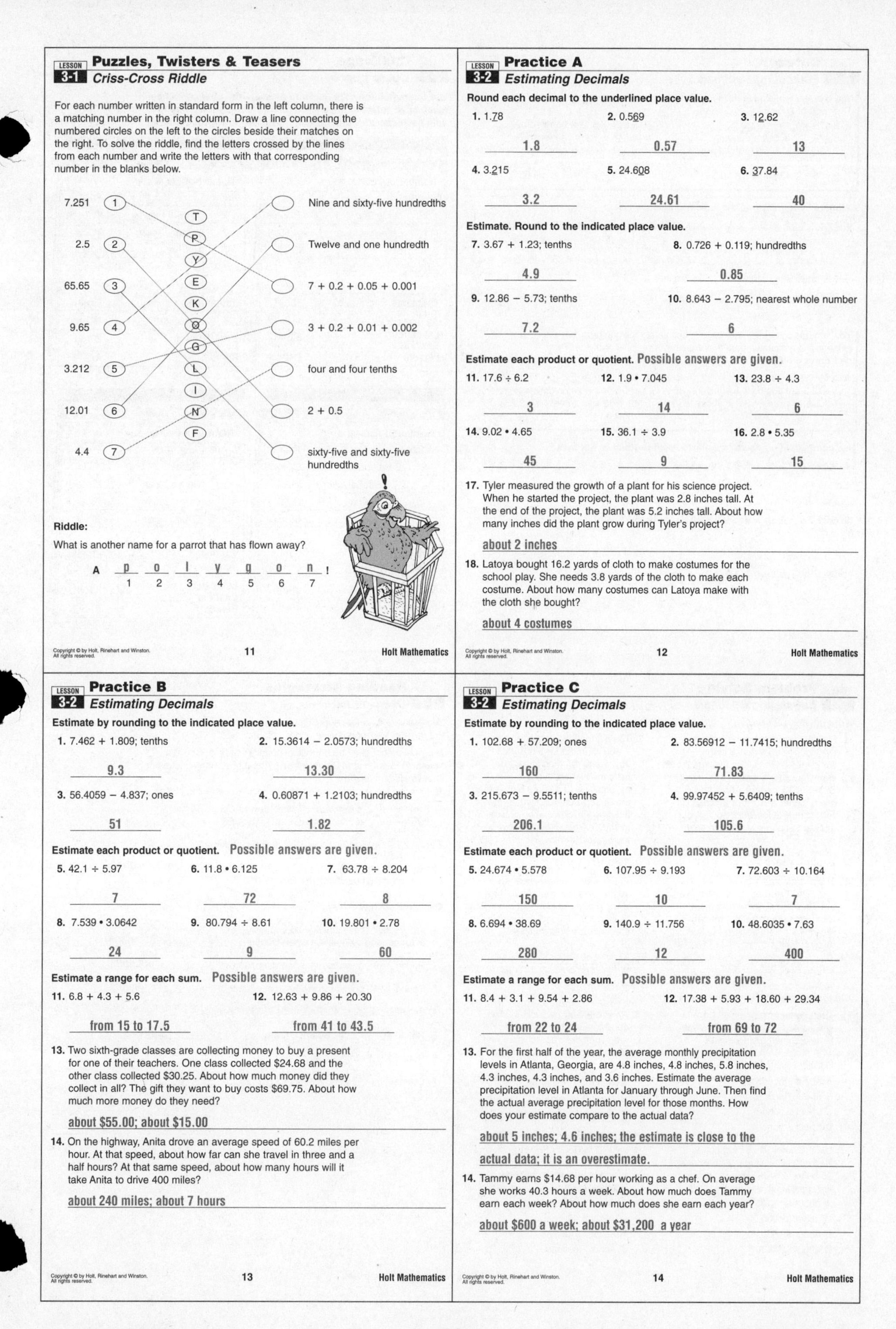

LESSON 3-1

Puzzles, Twisters & Teasers

Criss-Cross Riddle

For each number written in standard form in the left column, there is a matching number in the right column. Draw a line connecting the numbered circles on the left to the circles beside their matches on the right. To solve the riddle, find the letters crossed by the lines from each number and write the letters with that corresponding number in the blanks below.

Riddle:

What is another name for a parrot that has flown away?

A p o l y g o n !

1 2 3 4 5 6 7

 11 Holt Mathematics

LESSON 3-2

Practice A

Estimating Decimals

Round each decimal to the underlined place value.

1. 1.78 — 1.8
2. 0.569 — 0.57
3. 12.62 — 13
4. 3.215 — 3.2
5. 24.608 — 24.61
6. 37.84 — 40

Estimate. Round to the indicated place value.

7. 3.67 + 1.23; tenths — 4.9
8. 0.726 + 0.119; hundredths — 0.85
9. 12.86 − 5.73; tenths — 7.2
10. 8.643 − 2.795; nearest whole number — 6

Estimate each product or quotient. Possible answers are given.

11. 17.6 ÷ 6.2 — 3
12. 1.9 • 7.045 — 14
13. 23.8 ÷ 4.3 — 6
14. 9.02 • 4.65 — 45
15. 36.1 ÷ 3.9 — 9
16. 2.8 • 5.35 — 15

17. Tyler measured the growth of a plant for his science project. When he started the project, the plant was 2.8 inches tall. At the end of the project, the plant was 5.2 inches tall. About how many inches did the plant grow during Tyler's project?

about 2 inches

18. Latoya bought 16.2 yards of cloth to make costumes for the school play. She needs 3.8 yards of the cloth to make each costume. About how many costumes can Latoya make with the cloth she bought?

about 4 costumes

 12 Holt Mathematics

LESSON 3-2

Practice B

Estimating Decimals

Estimate by rounding to the indicated place value.

1. 7.462 + 1.809; tenths — 9.3
2. 15.3614 − 2.0573; hundredths — 13.30
3. 56.4059 − 4.837; ones — 51
4. 0.60871 + 1.2103; hundredths — 1.82

Estimate each product or quotient. Possible answers are given.

5. 42.1 ÷ 5.97 — 7
6. 11.8 • 6.125 — 72
7. 63.78 ÷ 8.204 — 8
8. 7.539 • 3.0642 — 24
9. 80.794 ÷ 8.61 — 9
10. 19.801 • 2.78 — 60

Estimate a range for each sum. Possible answers are given.

11. 6.8 + 4.3 + 5.6 — from 15 to 17.5
12. 12.63 + 9.86 + 20.30 — from 41 to 43.5

13. Two sixth-grade classes are collecting money to buy a present for one of their teachers. One class collected $24.68 and the other class collected $30.25. About how much money did they collect in all? The gift they want to buy costs $69.75. About how much more money do they need?

about $55.00; about $15.00

14. On the highway, Anita drove an average speed of 60.2 miles per hour. At that speed, about how far can she travel in three and a half hours? At that same speed, about how many hours will it take Anita to drive 400 miles?

about 240 miles; about 7 hours

 13 Holt Mathematics

LESSON 3-2

Practice C

Estimating Decimals

Estimate by rounding to the indicated place value.

1. 102.68 + 57.209; ones — 160
2. 83.56912 − 11.7415; hundredths — 71.83
3. 215.673 − 9.5511; tenths — 206.1
4. 99.97452 + 5.6409; tenths — 105.6

Estimate each product or quotient. Possible answers are given.

5. 24.674 • 5.578 — 150
6. 107.95 ÷ 9.193 — 10
7. 72.603 ÷ 10.164 — 7
8. 6.694 • 38.69 — 280
9. 140.9 ÷ 11.756 — 12
10. 48.6035 • 7.63 — 400

Estimate a range for each sum. Possible answers are given.

11. 8.4 + 3.1 + 9.54 + 2.86 — from 22 to 24
12. 17.38 + 5.93 + 18.60 + 29.34 — from 69 to 72

13. For the first half of the year, the average monthly precipitation levels in Atlanta, Georgia, are 4.8 inches, 4.8 inches, 5.8 inches, 4.3 inches, 4.3 inches, and 3.6 inches. Estimate the average precipitation level in Atlanta for January through June. Then find the actual average precipitation level for those months. How does your estimate compare to the actual data?

about 5 inches; 4.6 inches; the estimate is close to the actual data; it is an overestimate.

14. Tammy earns $14.68 per hour working as a chef. On average she works 40.3 hours a week. About how much does Tammy earn each week? About how much does she earn each year?

about $600 a week; about $31,200 a year

 14 Holt Mathematics

LESSON 3-2
Reteach
Estimating Decimals

You can use rounding to estimate. Round to the indicated place value. Then add or subtract.

A. 3.478 + 7.136; tenths

3.478 — 7 ≥ 5, so round up — 3.5
7.136 — 3 < 5, so round down — +7.1
10.6

3.478 + 7.136 is about 10.6.

B. 12.848 − 6.124; hundredths

12.848 — 8 ≥ 5, so round up — 12.85
6.124 — 4 < 5, so round down — −6.12
6.73

12.848 − 6.124 is about 6.73.

Estimate by rounding to the indicated place value.

1. 1.04 + 9.37; tenths

1.04 rounds to 1.0
9.37 rounds to 9.4
estimate 10.4

2. 2.17 + 3.56; tenths

2.17 rounds to 2.2
3.56 rounds to 3.6
estimate 5.8

3. 6.753 − 4.245; hundredths

6.753 rounds to 6.75
4.255 rounds to 4.26
estimate 2.49

You can use compatible numbers to estimate. Pick numbers that are close to the actual numbers that are easy to multiply or divide. Then multiply or divide.

A. 4.6 • 3.2
5 and 3 are compatible numbers.
5 • 3 = 15, so 4.6 • 3.2 is about 15.

B. 48.3 ÷ 13.2
48 and 12 are compatible numbers.
48 ÷ 12 = 4, so 48.3 ÷ 13.2 is about 4.

Use compatible numbers to estimate each product or quotient.

4. 9.4 • 5.6 — 54
5. 7.25 • 10.84 — 77
6. 84.8 ÷ 3.9 — 21
7. 21.9 ÷ 3.1 — 7
8. 8.3 • 7.6 — 64
9. 55.7 ÷ 6.9 — 8
10. 5.57 ÷ 2.7 — 2
11. 6.729 • 9.8 — 70

Possible answers are given.

LESSON 3-2
Challenge
Out to Lunch

Use the restaurant bills below to estimate the total cost of each meal. Then estimate the amount each person should pay to split each check evenly. Possible answers are given.

Number of People: 2

Quantity	Item	Price
2	Large Soda	$0.75 each
1	Cheeseburger	$4.55
1	BLT	$3.25
Estimated Total:		about $10.00
Estimated Cost Per Person:		about $5.00

Number of People: 4

Quantity	Item	Price
2	Slice of Pie	$1.45 each
1	Brownie	$1.25
3	Hot Tea	$0.60 each
Estimated Total:		about $6.00
Estimated Cost Per Person:		about $1.50

Number of People: 4

Quantity	Item	Price
2	Bowl of Soup	$2.89 each
1	Chicken Sandwich	$4.95
1	Chef Salad	$3.25
4	Coffee	$0.50 each
Estimated Total:		about $16.00
Estimated Cost Per Person:		about $4.00

Number of People: 5

Quantity	Item	Price
1	Pizza	$14.95
5	House Salad	$2.85 each
4	Large Soda	$0.75 each
1	Small Soda	$0.55
Estimated Total:		about $35.00
Estimated Cost Per Person:		about $7.00

LESSON 3-2
Problem Solving
Estimating Decimals

Write the correct answer. Possible answers are given.

1. Men in Iceland have the highest average life expectancy in the world—76.8 years. The average life expectancy for a man in the United States is 73.1 years. About how much higher is a man's average life expectancy in Iceland? Round your answer to the nearest whole year.

about 4 years

2. The average life expectancy for a woman in the United States is 79.1 years. Women in Japan have the highest average life expectancy—3.4 years higher than the United States. Estimate the average life expectancy of women in Japan. Round your answer to the nearest whole year.

about 82 years

3. There are about 1.6093 kilometers in one mile. There are 26.2 miles in a marathon race. About how many kilometers are there in a marathon race? Round your answer to the nearest tenths.

about 41.9 kilometers

4. At top speed, a hornet can fly 13.39 miles per hour. About how many hours would it take a hornet to fly 65 miles? Round your answer to the nearest whole number.

about 5 hours

Circle the letter of the correct answer.

5. The average male human brain weighs 49.7 ounces. The average female human brain weighs 44.6 ounces. What is the difference in their weights?

A about 95 ounces
B about 7 ounces
(C) about 5 ounces
D about 3 ounces

6. An official hockey puck is 2.54 centimeters thick. About how thick are two hockey pucks when one is placed on top of the other?

F about 4 centimeters
G about 4.2 centimeters
(H) about 5 centimeters
J about 5.2 centimeters

7. Lydia earned $9.75 per hour as a lifeguard last summer. She worked 25 hours a week. About how much did she earn in 8 weeks?

A about $250.00
(B) about $2,000.00
C about $2,500.00
D about $200.00

8. Brent mixed 4.5 gallons of blue paint with 1.7 gallons of white paint and 2.4 gallons of red paint to make a light purple paint. About how many gallons of purple paint did he make?

(F) about 9 gallons
G about 8 gallons
H about 10 gallons
J about 7 gallons

LESSON 3-2
Reading Strategies
Use Context

You estimate to get an approximate answer. Rounding decimals to the nearest whole number is one way to estimate.

Mike's mom bought 3.28 pounds of cheddar cheese. She also bought 2.75 pounds of Swiss cheese. About how many pounds of cheese did she buy?

To round to the nearest whole number, look at the tenths place.

3.28 ← 2 is less than 5; round down to 3.
+2.75 ← 7 is greater than 5; round up to 3.

3.28 + 2.75 rounded to the nearest whole number is:

3 + 3 = 6 pounds of cheese.

Complete each problem.

1. Which decimal place value do you look at to round to the nearest whole number? tenths place

2. Round 34.67 pounds to the nearest pound. 35 pounds

3. Round 42.19 pounds to the nearest pound. 42 pounds

4. Estimate this sum: 42.19 pounds + 34.67 pounds. 42 + 35 = 77

5. Round $54.14 to the nearest dollar. $54

6. Round $21.54 to the nearest dollar. $22

7. Estimate the difference: $54.14 − $21.54. $54 − $22 = $32

LESSON **3-2**

Puzzles, Twisters & Teasers

Secret Message

Recently you've noticed that your neighbor Slippery Larry has been extremely forgetful when he leaves his house. He keeps leaving, then returning to get all of the things he forgot. One day you decided to keep track of how far he had gone by measuring how many feet he walked down his driveway before returning to the house. Here's what you found:

Item	Distance	Item	Distance
1. Keys	6.435	8. Soccer Ball	13.575
2. Wallet	4.909	9. Laptop	18.249
3. Jacket	4.899	10. Suspenders	13.59
4. Pants	14.642	11. Rope	1.59
5. Watch	12.94	12. Gloves	14.253
6. Briefcase	4.851	13. Hat	18.161
7. Pen	1.491		

It just so happens that Slippery Larry was on the news later that day. He is actually a criminal communicating to his criminal buddies through a code system. Seems his friends watch how far he travels down his driveway and the distances translate into the following code:

1.6	A	19.0	H	9.4	O	13.0	V
3.6	B	21.2	I	8.8	P	13.8	W
8.4	C	14.8	J	9.6	Q	1.4	X
4.4	D	9.2	K	10.8	R	1.2	Y
5.0	E	2.6	L	18.2	S	5.2	Z
6.8	F	6.4	M	13.6	T		
15.6	G	14.6	N	14.4	U		

Since your measurements are more exact than Slippery Larry's and his friends', you'll have to round each of the distances to get the total distance traveled. Translate the code into letters and fill in the blanks to discover where you can catch Slippery Larry!

M E E T   M E
1 2 2 8   1 2

A T   S T A N ' S
11 8   13 8 11 4 13

LESSON **3-3**

Practice A

Adding and Subtracting Decimals

Find each sum or difference.

1. 1.5 + 2.3   3.8
2. 6.5 + 1.4   7.9
3. 8.9 − 5.1   3.8
4. 12.6 − 3.4   9.2
5. 8.16 − 7.02   1.14
6. 7.25 + 8.75   16.00
7. 11.4 + 8.6   20.0
8. 16.5 − 4.3   12.2
9. 9.55 − 1.2   8.35
10. 25.6 + 5.1   30.7
11. 8.9 + 3.05   11.95
12. 10.64 − 8.5   2.14

Circle the letter of the correct answer.

13. If $x = 2.3$, what is the value of the expression $5.4 + x$?
A 3.1
(B) 7.7
C 7.1
D 3.7

14. If $a = 4.2$, what is the value of the expression $8.7 - a$?
F 12.9
G 4.9
(H) 4.5
J 12.5

15. If $m = 1.9$, what is the value of the expression $m + 4.2$?
A 2.3
B 2.2
(C) 6.1
D 7.1

16. If $y = 5.9$, what is the value of the expression $7.2 - y$?
(F) 1.3
G 1.7
H 13.3
J 13.1

17. Marcus is 1.5 meters tall. His sister, Carol, is 0.1 meter taller than Marcus. Their father is 0.2 meter taller than Carol. How tall is Carol? How tall is their father?
1.6 meters; 1.8 meters

18. Jennifer brought $14.75 to the baseball game. She spent $3.45 for a hot dog and soda. How much money does she have left?
$11.30

LESSON **3-3**

Practice B

Adding and Subtracting Decimals

Find each sum or difference.

1. 8.9 + 2.4   11.3
2. 12.7 − 9.6   3.1
3. 18.35 − 4.16   14.19
4. 7.21 + 11.6   18.81
5. 0.975 + 3.8   4.775
6. 20.66 − 9.1   11.56

7. Tiffany's job requires a lot of driving. How many miles did she travel during the month of February? 720.78 miles

Miles Tiffany Traveled

Week	1	2	3	4
Miles	210.05	195.18	150.25	165.30

8. Shelly babysits after school and on the weekends. How much did she earn in all for the month of April? $722.00

Shelly's Earnings for April

Week	1	2	3	4
Earnings	$120.50	$180.75	$205.25	$215.50

Evaluate 5.6 − *a* for each value of *a*.

9. $a = 3.7$ 1.9
10. $a = 0.5$ 5.1
11. $a = 2.8$ 2.8
12. $a = 1.42$ 4.18
13. $a = 0.16$ 5.44
14. $a = 3.75$ 1.85

15. Allen bought a box of envelopes for $2.79 and a pack of paper for $4.50. He paid with a $10 bill. How much change should be receive?
$2.71

16. From a bolt of cloth measuring 25.60 yards, Tina cut a 6.8-yard piece and an 11.9-yard piece. How much material is left on the bolt?
6.9 yards

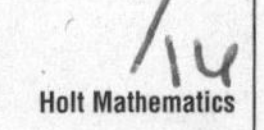

LESSON **3-3**

Practice C

Adding and Subtracting Decimals

Find each sum or difference.

1. 7.9 + 1.38   9.28
2. 100 − 65.9   34.1
3. 204.965 + 55.8   260.765
4. 63.057 − 18.45   44.607
5. 0.5541 − 0.09   0.4641
6. 11.2398 + 8.9 + 2.7   22.8398
7. 0.22 + 15.607 + 9.7   25.527
8. 20 − 3.78 − 0.64   15.58
9. 1.9 + 0.25 + 9.4 + 0.5   12.05

Evaluate each expression.

10. $39.702 - a$ for $a = 0.9$ 38.802
11. $x + 1.064$ for $x = 28.5$ 29.564
12. $50.02 - p$ for $p = 0.99$ 49.03
13. $w - 7.08$ for $w = 100$ 92.92
14. $37.62 + t$ for $t = 8.084$ 45.704
15. $u + 2.7$ for $u = 0.5046$ 3.2046

Write the missing digit in each problem.

16. 7.089 + 2.□13 = 9.502   4
17. 16.594 − □.175 = 11.419   5
18. 6.2□67 + 9.75 = 15.9867   3
19. 0.6□9 − 0.458 = 0.221   7
20. 238.793 + 19.5□2 = 258.305   1
21. 100.□5 − 19.99 = 80.46   4

22. Italian Delight sells three sizes of pizzas at different prices. If you buy all three pizzas, it costs a total of $46.24. A medium pizza costs $15.75, and a large costs $17.50. How much does a small pizza cost?
$12.99

23. Brent has three sheets of plywood that are each 6.85 feet long. He cut a 3.4-foot piece from one sheet and a 0.5-foot piece from the other. How many feet of plywood does he have left in all?
16.65 feet

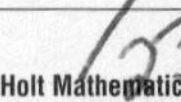

LESSON 3-3
Reteach
Adding and Subtracting Decimals

You can use a place-value chart to help you add and subtract decimals.

Add 1.4 and 0.9.

	Tens	Ones	Tenths	Hundredths	Thousandths
		1	.4		
+		0	.9		

So, 1.4 + 0.9 = 2.3.

Subtract 2.4 from 3.1.

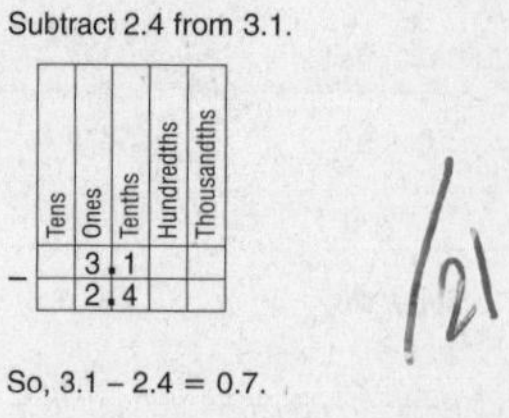

	Tens	Ones	Tenths	Hundredths	Thousandths
		3	.1		
−		2	.4		

So, 3.1 − 2.4 = 0.7.

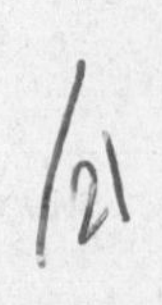

Find each sum or difference.

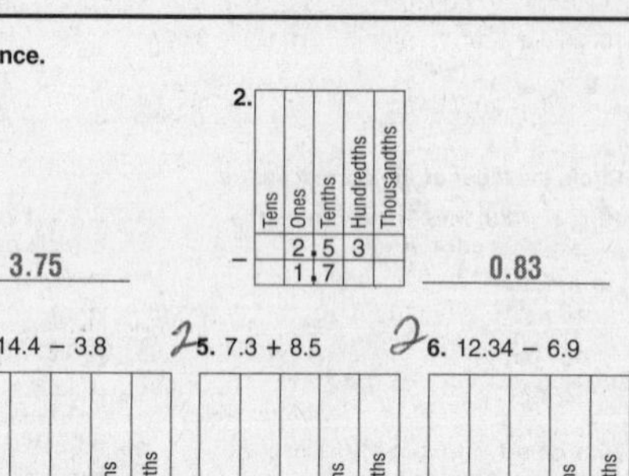

1.

	Tens	Ones	Tenths	Hundredths	Thousandths
		2	.6		
+		1	.1	5	

3.75

2.

	Tens	Ones	Tenths	Hundredths	Thousandths
		2	.5	3	
−		1	.7		

0.83

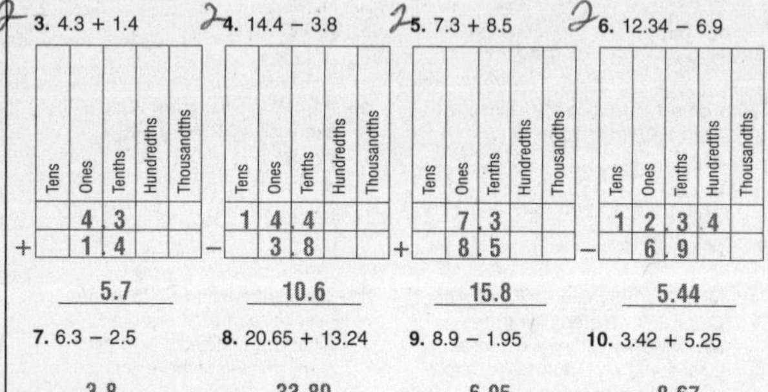

3. 4.3 + 1.4

	Tens	Ones	Tenths	Hundredths	Thousandths
		4	.3		
+		1	.4		

5.7

4. 14.4 − 3.8

	Tens	Ones	Tenths	Hundredths	Thousandths
	1	4	.4		
−		3	.8		

10.6

5. 7.3 + 8.5

	Tens	Ones	Tenths	Hundredths	Thousandths
		7	.3		
+		8	.5		

15.8

6. 12.34 − 6.9

	Tens	Ones	Tenths	Hundredths	Thousandths
	1	2	.3	4	
−		6	.9		

5.44

7. 6.3 − 2.5   3.8

8. 20.65 + 13.24   33.89

9. 8.9 − 1.95   6.95

10. 3.42 + 5.25   8.67

/14

LESSON 3-3
Challenge
A Penny Saved Is a Penny Earned

Next to each bank, describe three different coin combinations that equal the amount of money it holds. For each combination, use at least one quarter, one dime, one nickel, and one penny.

Possible combinations are given.

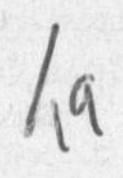

1.

2 quarters, 3 dimes, 1 nickel, 4 pennies

2 quarters, 2 dimes, 2 nickels, 9 pennies

1 quarter, 5 dimes, 2 nickels, 4 pennies

2.

4 quarters, 2 dimes, 1 nickel, 3 pennies

2 quarters, 4 dimes, 5 nickels, 13 pennies

3 quarters, 4 dimes, 2 nickels, 3 pennies

3.

1 quarter, 3 dimes, 1 nickel, 5 pennies

1 quarter, 1 dime, 4 nickels, 10 pennies

1 quarter, 2 dimes, 3 nickels, 5 pennies

4.

8 quarters, 2 dimes, 1 nickel, 5 pennies

4 quarters, 10 dimes, 4 nickels, 10 pennies

6 quarters, 3 dimes, 4 nickels, 30 pennies

/12

LESSON 3-3
Problem Solving
Adding and Subtracting Decimals

Use the table to answer the questions.

Busiest Ports in the United States

Port	Imports Per Year (millions of tons)	Exports Per Year (millions of tons)
South Louisiana, LA	30.6	57.42
Houston, TX	75.12	33.43
New York, NY & NJ	53.52	8.03
New Orleans, LA	26.38	21.73
Corpus Christi, TX	52.6	7.64

1. How many more tons of imports than exports does the Port of New Orleans handle each year?

   4.65 million tons

2. How many tons of imports and exports are shipped through the port of Houston, Texas, each year in all?

   108.55 million tons

Circle the letter of the correct answer.

3. Which port ships 0.39 more tons of exports each year than the port at Corpus Christi, Texas?
   - A Houston
   - (B) NY & NJ
   - C New Orleans
   - D South Louisiana

4. What is the difference between the imports and exports shipped in and out of Corpus Christi's port each year?
   - F 45.04 million tons
   - G 44.94 million tons
   - (H) 44.96 million tons
   - J 44.06 million tons

5. What is the total amount of imports shipped into the nation's 5 busiest ports each year?
   - (A) 238.22 million tons
   - B 366.47 million tons
   - C 128.25 million tons
   - D 109.97 million tons

6. What is the total amount of exports shipped out of the nation's 5 busiest ports each year?
   - F 366.47 million tons
   - (G) 128.25 million tons
   - H 109.97 million tons
   - J 238.22 million tons

LESSON 3-3
Reading Strategies
Use an Organizer

Writing decimals in a place-value grid helps you line up decimal points to add or subtract decimals.

	Ones	Tenths	Hundredths
	1	.4	0
	5	.3	8
+	2	.7	0
	9	.4	8

1.40
5.38
+ 2.70
9.48

Add zeros as place holders.

Place decimal point in answer.

28.05
− 6.30
21.75

	Tens	Ones	Tenths	Hundredths
	2	8	.0	5
−		6	.3	0
	2	1	.7	5

1. How does the place-value grid help you add or subtract?

   It helps you line up decimal points.

2. Place these numbers on the place-value grid below: 3.25, 1.06, 2.9.

	Ones	Tenths	Hundredths
	3	.2	5
	1	.0	6
+	2	.9	0

3. Add the numbers on the place-value grid. What is the sum?

   7.21

4. Place this problem on the place-value grid below: 23.82 − 7.2.

	Tens	Ones	Tenths	Hundredths
	2	3	.8	2
−		7	.2	0

5. Subtract the numbers on the place-value grid. What is the difference?

   16.62

6. For which numbers did you add zero as a place holder?

   2.9 written as 2.90; 7.2 written as 7.20

/7

LESSON 3-3

Puzzles, Twisters & Teasers

Crazy Pete's Letter Shop

What's the quickest way to double your money?

To solve this riddle, you need to buy your letters from Crazy Pete.

Right now, you know three letters from the first word: **L, F, O.** You also know three letters from the second word: **F, A, H.**

To use those six letters in the riddle, you need to buy them from Crazy Pete.

You have $15.00 to spend on each word, but you know you need to buy two more letters to complete the riddle—one for each word.

Crazy Pete tells you a secret: "You have exactly enough money left after buying the first three letters of each word to purchase the missing letter."

Using the price table at right, add the cost of the three letters you need to buy for the first word. Subtract this total from $15.00 to find the cost of the missing letter.

Then do it again for the second word.

Letter	Price
A	$1.50
B	$5.43
C	$6.08
D	$4.13
E	$1.50
F	$6.08
G	$4.79
H	$2.83
I	$1.50
J	$6.08
K	$4.79
L	$4.59
M	$5.43
N	$3.29
O	$0.20
P	$4.79
Q	$7.87
R	$3.29
S	$3.29
T	$3.29
U	$1.50
V	$4.79
W	$5.43
X	$7.87
Y	$5.43
Z	$7.87

First Word Cost (letters **L, F, O**): 10.87

Remaining money: 4.13

Missing Letter: D

Second Word Cost (letters **F, A, H**): 10.41

Remaining money: 4.59

Final Letter: L

Unscramble the letters of each word to solve the riddle!

F O L D it in

H A L F !

LESSON 3-4

Practice A

Scientific Notation

Find each product.

1. $267 \cdot 100$ — 26,700
2. $38.1 \cdot 100$ — 3,810
3. $1.92 \cdot 100$ — 192

Circle the letter of the correct answer.

4. Which of the following shows 85,000 written in scientific notation?
   A $8.5 \cdot 10^3$
   (B) $8.5 \cdot 10^4$
   C $8.5 \cdot 10^5$
   D $8.5 \cdot 10^6$

5. Which of the following shows $3.67 \cdot 10^5$ written in standard form?
   F 3,670
   G 36,700
   (H) 367,000
   J 3,670,000

Fill in the blanks to make each equation true.

6. $1{,}200 = 1.2 \cdot 10^{\_}$ — 3
7. $25{,}000 = 2.5 \cdot 10^{\_}$ — 4
8. $580 = 5.8 \cdot 10^{\_}$ — 2
9. $470{,}000 = \_\_ \cdot 10^5$ — 4.7
10. $6{,}580 = \_\_ \cdot 10^3$ — 6.58
11. $8{,}900{,}000 = \_\_ \cdot 10^6$ — 8.9

Write each number in standard form.

12. $3.4 \cdot 10^2$ — 340
13. $7.9 \cdot 10^4$ — 79,000
14. $1.75 \cdot 10^3$ — 1,750
15. $1.24 \cdot 10^5$ — 124,000
16. $9.6 \cdot 10^5$ — 960,000
17. $1.28 \cdot 10^6$ — 1,280,000

18. African elephants are the largest land mammals. The average African elephant weighs 11,000 pounds. Write this weight in scientific notation.

    $1.1 \cdot 10^4$ pounds

LESSON 3-4

Practice B

Scientific Notation

Find each product.

1. $345 \cdot 100$ — 34,500
2. $65.2 \cdot 100$ — 6,520
3. $1.84 \cdot 1{,}000$ — 1,840

Write each number in scientific notation.

4. 16,700 — $1.67 \cdot 10^4$
5. 4,680 — $4.68 \cdot 10^3$
6. 58,340,000 — $5.834 \cdot 10^7$

Write each number in standard form.

7. $3.25 \cdot 10^4$ — 32,500
8. $7.08 \cdot 10^6$ — 7,080,000
9. $1.209 \cdot 10^7$ — 12,090,000
10. $6.8 \cdot 10^8$ — 680,000,000
11. $0.51 \cdot 10^5$ — 51,000
12. $0.006 \cdot 10^3$ — 6

Identify the answer choice that is *not* equal to the given number.

13. 356,000
    A 300,000 + 56,000
    B $3.56 \cdot 10^5$
    (C) $3.56 \cdot 10^4$

14. $1.28 \cdot 10^6$
    (A) 100,000 + 28,000
    B 1,280,000
    C $12.8 \cdot 10^5$

15. 1,659,000
    A 1,600,000 + 59,000
    B $1.659 \cdot 10^6$
    (C) $16.59 \cdot 10^6$

16. $0.074 \cdot 10^3$
    A 70.0 + 4.0
    (B) $7.4 \cdot 10^5$
    C $7.4 \cdot 10^1$

17. In 2000, the population of Pennsylvania was 12,281,054. Round this figure to the nearest hundred thousand. Then write that number in scientific notation.

    12,300,000; $1.23 \cdot 10^7$

18. In 2000, the population of North Carolina was about $8.05 \cdot 10^6$, and the population of South Carolina was about $4.01 \cdot 10^6$. Write the combined populations of these two states in standard form.

    12,060,000

LESSON 3-4

Practice C

Scientific Notation

Find each product.

1. $1.67 \times 1{,}000$ — 1,670
2. 93.6×100 — 9,360
3. $3.55 \times 10{,}000$ — 35,500

Write each number in scientific notation.

4. 6,389,000 — $6.389 \cdot 10^6$
5. 105,200,000 — $1.052 \cdot 10^8$
6. 152 million — $1.52 \cdot 10^8$

Write each number in standard form.

7. $1.5089 \cdot 10^4$ — 15,089
8. $2.516 \cdot 10^8$ — 251,600,000
9. $1.7711 \cdot 10^7$ — 17,711,000
10. $3.9604 \cdot 10^6$ — 3,960,400
11. $0.284 \cdot 10^4$ — 2,840
12. $0.0869 \cdot 10^2$ — 8.69

Write each measurement using scientific notation.

13. 250 km = $2.5 \cdot 10^5$ m
14. 0.065 kg = $6.5 \cdot 10^1$ g
15. 89 L = $8.9 \cdot 10^4$ mL
16. 1,540 km = $1.54 \cdot 10^8$ cm
17. 0.73 m = $7.3 \cdot 10^2$ mm
18. 10,240 kg = $1.024 \cdot 10^7$ g

19. In a recent count, 147,171,000 people in the United States owned cars. In the same year, $4.268 \cdot 10^7$ people in Japan owned cars. In which country did more people own cars? How many more?

    the United States; 104,491,000 more people

20. On average, about $1.1 \cdot 10^9$ passengers use the New York City subway system each year. About 1,170,000,000 passengers use the Paris subway each year. How many passengers use those two subways each year?

    2,270,000,000 passengers

LESSON **3-4**

Reteach

Scientific Notation

Scientific notation expresses a large number as the product of a number between one and ten and a power of ten.

To write 3,400 in scientific notation, move the decimal point to the left until the number falls between 1 and 10.

3,400 $1 < 3 < 10$, so move the decimal point 3 places to the left.

$3{,}400 = 3.4 \cdot 10^3$ The number of times you move the decimal point left is the power of ten.

Express each number in scientific notation.

1. 175,000 — $1.75 \cdot 10^5$
2. 298 — $2.98 \cdot 10^2$
3. 5,764 — $5.764 \cdot 10^3$
4. 83 — $8.3 \cdot 10^1$
5. 40,300 — $4.03 \cdot 10^4$
6. 2,000,000 — $2 \cdot 10^6$
7. 51,010 — $5.101 \cdot 10^4$
8. 190,025 — $1.90025 \cdot 10^5$

You can express numbers written in scientific notation in standard form.

The power of ten tells you how many places to move the decimal point to the right.

$3.2 \cdot 10^4 = 32{,}000$ To write $3.2 \cdot 10^4$ in standard form, move the decimal point 4 places to the right.

Write each number in standard form.

9. $5.62 \cdot 10^3$ — 5,620
10. $7.238 \cdot 10^2$ — 723.8
11. $9.9 \cdot 10^5$ — 990,000
12. $6.53 \cdot 10^1$ — 65.3
13. $5.36 \cdot 10^4$ — 53,600
14. $2.4 \cdot 10^2$ — 240
15. $4.35 \cdot 10^3$ — 4,350
16. $8 \cdot 10^5$ — 800,000
17. $1 \cdot 10^4$ — 10,000
18. $2.03 \cdot 10^3$ — 2030
19. $1.12 \cdot 10^2$ — 112
20. $3.002 \cdot 10^6$ — 3,002,000

/20

LESSON **3-4**

Challenge

The Solar System

Write the average distance from the Sun in standard form. Then use the distances to label our solar system shown below.

/31

		Average Distance From the Sun (mi)	
		Scientific Notation	Standard Form
1.	Earth	$9.29 \cdot 10^7$	92,900,000
2.	Jupiter	$4.836 \cdot 10^8$	483,600,000
3.	Mars	$1.416 \cdot 10^8$	141,600,000
4.	Mercury	$3.6 \cdot 10^7$	36,000,000
5.	Neptune	$2.794 \cdot 10^9$	2,794,000,000
6.	Pluto	$3.675 \cdot 10^9$	3,675,000,000
7.	Saturn	$8.87 \cdot 10^8$	887,000,000
8.	Uranus	$1.784 \cdot 10^9$	1,784,000,000
9.	Venus	$6.72 \cdot 10^7$	67,200,000

Asteroid Belt
Saturn
Pluto
Neptune
Mercury
Mars
Earth
Uranus
Venus
Jupiter
Sun

/19

LESSON **3-4**

Problem Solving

Scientific Notation

Write the correct answer.

1. The closest comet to approach Earth was called Lexell. On July 1, 1770, Lexell was observed about 874,200 miles from Earth's surface. Write this distance in scientific notation.

   $8.742 \cdot 10^5$

2. Scientists estimate that it would take $1.4 \cdot 10^{10}$ years for light from the edge of our universe to reach Earth. How many years is that written in standard form?

   14,000,000,000 years

3. In the United States, about 229,000,000 people speak English. About 18,000,000 people speak English in Canada. Write in scientific notation the total number of English speaking people in the United States and Canada.

   $2.47 \cdot 10^8$ people

4. South Africa is the top gold-producing country in the world. Each year it produces $4.688 \cdot 10^8$ tons of gold! Written in standard form, how many tons of gold does South African produce each year?

   468,800,000 tons

Circle the letter of the correct answer.

5. About $3.012 \cdot 10^6$ people visit Yellowstone National Park each year. What is that figure written in standard form?
   - A 30,120,000 people
   - (B) 3,012,000 people
   - C 301,200 people
   - D 30,120 people

6. In 2000, farmers in Iowa grew 1,740,000 bushels of corn. What is this amount written in scientific notation?
   - F $1.7 \cdot 10^5$
   - G $1.74 \cdot 10^5$
   - (H) $1.74 \cdot 10^6$
   - J $1.74 \cdot 10^7$

7. The temperature at the core of the Sun reaches 27,720,000°F. What is this temperature written in scientific notation?
   - A $2.7 \cdot 10^7$
   - B $2.72 \cdot 10^7$
   - C $2.772 \cdot 10^6$
   - (D) $2.772 \cdot 10^7$

8. Your body is constantly producing red blood cells—about $1.73 \cdot 10^{11}$ cells a day. How many blood cells is that written in standard form?
   - F 173,000,000 cells
   - G 17,300,000,000 cells
   - (H) 173,000,000,000 cells
   - J 1,730,000,000,000 cells

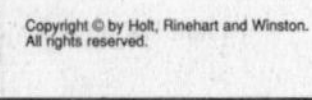

LESSON **3-4**

Reading Strategies

Use a Graphic Organizer

This chart helps you see the ways large numbers can be written.

Scientific Notation
- A number between 1 and 10 multiplied by a power of 10
  - 2.5×10^6
  - 7×10^5

Ways to Write Large Numbers

Standard Form	Words and Symbols
• All place values are shown. 2,500,000 700,000	• Use numbers and words. 2 million, 500 thousand; 2.5 million 700 thousand

Use the graphic organizer to answer Exercises 1–2.

1. Which way to write large numbers shows every place value?

   standard form

2. Which way to write large numbers uses a power of 10?

   scientific notation

Identify how each large number is written. Write "scientific notation," "standard form," or "words and symbols".

3. 8,296,000

   standard form

4. 3.6 million

   words and symbols

5. 2.9×10^5

   scientific notation

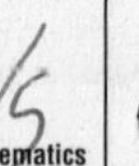

LESSON 3-4
Puzzles, Twisters & Teasers
The Scientific Two-Step

There are two simple steps to do the Scientific Two-Step and answer the riddle:

1. For each number in scientific notation in the left column, find the same number in standard notation in the key on the right.
2. Fill in the letter associated with the standard notation answer to solve the riddle.

		Key:	
1	2.458×10^1	12,760	E
2	1.276×10^8	127,600,000	O
3	2.458×10^2	24,580,000	D
4	1.276×10^6	245.8	M
5	2.458×10^4	24,580	A
6	2.458×10^9	2,458	F
7	1.276×10^7	12,760,000	Y
8	2.458×10^3	1,276	L
9	2.458×10^5	24.58	C
10	1.276×10^3	245,800	O
11	2.458×10^7	1,276,000	P
12	1.276×10^4	2,458,000,000	N
13	1.276×10^2	127.6	D

Why did the people working at the blanket factory lose their jobs?

13 The c o m p a n y (1 2 3 4 5 6 7) f o l d e d (8 9 10 11 12 13)

/13

LESSON 3-5
Practice A
Multiplying Decimals

Find each product.

1. 0.4 × 0.2 — 0.08
2. 0.3 × 0.4 — 0.12
3. 1.2 × 0.5 — 0.6
4. 1.1 × 0.9 — 0.99
5. 2.5 × 0.5 — 1.25
6. 6.0 × 0.7 — 4.2
7. 0.4 • 0.5 — 0.2
8. 1.2 • 1.5 — 1.8
9. 1.7 • 0.3 — 0.51
10. 6.7 • 0.4 — 2.68
11. 9.6 • 0.2 — 1.92
12. 0.8 • 0.8 — 0.64

Evaluate 2*x* for each value of *x*.

13. $x = 0.1$ — 0.2
14. $x = 0.5$ — 1.0
15. $x = 0.9$ — 1.8
16. $x = 1.2$ — 2.4
17. $x = 1.7$ — 3.4
18. $x = 2.4$ — 4.8

19. Each box can hold 2.5 pounds of apples. How many pounds can 3 boxes hold?
 7.5 pounds

20. Each pie costs $5.60. How much will it cost to buy 2 pies?
 $11.20

/20

LESSON 3-5
Practice B
Multiplying Decimals

Find each product.

1. 0.7 × 0.3 — 0.21
2. 0.05 × 0.4 — 0.02
3. 8.0 × 0.02 — 0.16
4. 3.5 × 0.2 — 0.7
5. 12.1 × 0.01 — 0.121
6. 9.0 × 0.9 — 8.1
7. 0.04 • 0.58 — 0.0232
8. 2.15 • 1.5 — 3.225
9. 1.73 • 0.8 — 1.384
10. 6.017 • 2.0 — 12.034
11. 3.96 • 0.4 — 1.584
12. 0.7 • 0.009 — 0.0063

Evaluate 8*x* for each value of *x*.

13. $x = 0.5$ — 4
14. $x = 2.3$ — 18.4
15. $x = 0.74$ — 5.92
16. $x = 3.12$ — 24.96
17. $x = 0.587$ — 4.696
18. $x = 14.08$ — 112.64

19. The average mail carrier walks 4.8 kilometers in a workday. How far do most mail carriers walk in a 6-day week? There are 27 working days in July, so how far will a mail carrier walk in July?
 28.8 kilometers;
 129.6 kilometers

20. A deli charges $3.45 for a pound of turkey. If Tim wants to purchase 2.4 pounds, how much will it cost?
 $8.28

LESSON 3-5
Practice C
Multiplying Decimals

Find each product.

1. 9.86 × 0.3 — 2.958
2. 12.01 × 0.46 — 5.5246
3. 7.05 × 0.03 — 0.2115
4. 11.65 • 0.23 — 2.6795
5. 24.54 • 0.037 — 0.90798
6. 11.405 • 2.91 — 33.18855
7. 0.058 • 0.129 — 0.007482
8. 29.864 • 5.13 — 153.20232
9. 100.86 • 0.004 — 0.40344

Evaluate 17*x* for each value of *x*.

10. $x = 1.9$ — 32.3
11. $x = 0.005$ — 0.085
12. $x = 6.307$ — 107.219
13. $x = 11.215$ — 190.655
14. $x = 2.059$ — 35.003
15. $x = 75.844$ — 1,289.348

Evaluate.

16. $2.97n$ for $n = 1.8$ — 5.346
17. $12^2 + 1.9c$ for $c = 3.7$ — 151.03
18. $7^3 - 2x$ for $x = 0.54$ — 341.92
19. $1.6t + 3.056$ for $t = 2.09$ — 6.4

20. One year on Mercury is equal to 87.97 Earth days. One year on Pluto is three times the length of one Mercury year minus 16.21 days. How long is one year on Pluto?
 247.7 days

21. One year on Earth is equal to 365.30 days. One year on Mars is twice the length of one Earth year minus 43.6 days. How long is one year on Mars?
 687 days

/21

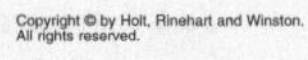

LESSON 3-5

Reteach

Multiplying Decimals

You can use a model to help you multiply a decimal by a whole number.

Find the product of 0.12 and 4, using a 10 by 10 grid.

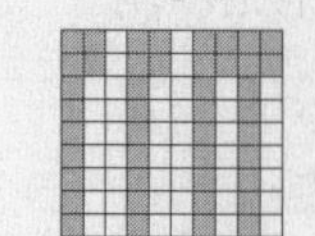

Shade 4 groups of 12 squares. Count the number of shaded squares. Since you have shaded 48 of the 100 squares, 0.12 • 4 = 0.48.

Find each product.

1. 0.23 • 3 **0.69**
2. 0.41 • 2 **0.82**
3. 0.011 • 5 **0.055**
4. 0.32 • 2 **0.64**
5. 0.15 • 3 **0.45**
6. 0.42 • 2 **0.84**
7. 0.04 • 8 **0.32**
8. 0.22 • 4 **0.88**

You can also use a model to help you multiply a decimal by a decimal.

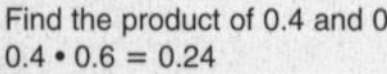

Find the product of 0.4 and 0.6.
0.4 • 0.6 = 0.24

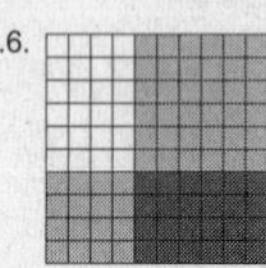

Find each product.

9. 0.2 • 0.8 **0.16**
10. 0.7 • 0.9 **0.63**
11. 0.5 • 0.5 **0.25**
12. 0.3 • 0.6 **0.18**
13. 0.5 • 0.2 **0.1**
14. 0.4 • 0.4 **0.16**
15. 0.1 • 0.9 **0.09**
16. 0.4 • 0.7 **0.28**

LESSON 3-5

Challenge

Decimal Growth

Use the growth rate for each plant below to find how much it will grow in 1 week.

Eucalyptus Tree	**Bristlecone Pine Tree**	**Trumpet Tree**
Growth Rate: 2.5 cm per day	**Growth Rate: 0.009 mm per day**	**Growth Rate: 0.28 in. per day**
17.5 cm	0.063 mm	1.96 in

Use the growth rate for each plant below to find how much it will grow in 0.25 day.

Oak Tree	**Lichens**	**Poplar Tree**
Growth Rate: 1.4 mm per day	**Growth Rate: 0.0025 mm per day**	**Growth Rate: 0.118 in. per day**
0.35 mm	0.000625 mm	0.0295 in.

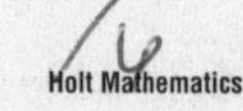

LESSON 3-5

Problem Solving

Multiplying Decimals

Use the table to answer the questions.

United States Minimum Wage

Year	Hourly Rate
1940	$0.30
1950	$0.75
1960	$1.00
1970	$1.60
1980	$3.10
1990	$3.80
2000	$5.15

1. At the minimum wage, how much did a person earn for a 40-hour workweek in 1950?

   $30.00

2. At the minimum wage, how much did a person earn for working 25 hours in 1970?

   $40.00

3. If you had a minimum-wage job in 1990, and worked 15 hours a week, how much would you have earned each week?

   $57.00

4. About how many times higher was the minimum wage in 1960 than in 1940?

   about 3 times

Circle the letter for the correct answer.

5. Ted's grandfather had a minimum-wage job in 1940. He worked 40 hours a week for the entire year. How much did Ted's grandfather earn in 1940?
   - A $12.00
   - (B) $624.00
   - C $642.00
   - D $6,240.00

6. Marci's mother had a minimum-wage job in 1980. She worked 12 hours a week. How much did Marci's mother earn each week?
   - F $3.72
   - G $37.00
   - H $37.10
   - (J) $37.20

7. Having one dollar in 1960 is equivalent to having $5.82 today. If you worked 40 hours a week in 1960 at minimum wage, how much would your weekly earnings be worth today?
   - A $40.00
   - B $5.82
   - (C) $232.80
   - D $2,328.00

8. In 2000, Cindy had a part-time job at a florist, where she earned minimum wage. She worked 18 hours each week for the whole year. How much did she earn from this job in 2000?
   - F $927.00
   - (G) $4,820.40
   - H $10,712.00
   - J $2,142.40

LESSON 3-5

Reading Strategies

Use a Visual Tool

Each grid shows 0.15 shaded.

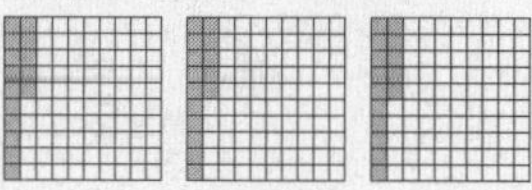

You can add the decimals to find how much of the grids are shaded. → 0.15 + 0.15 + 0.15 = 0.45

You can multiply 0.15 by 3. →

$$\begin{array}{r} \overset{1}{0.15} \\ \times\ 3 \\ \hline 0.45 \end{array}$$

Use these grids to complete the problems below.

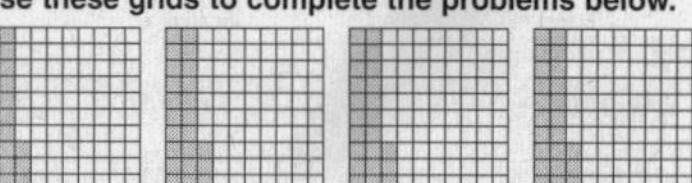

1. Shade 0.23 in each of the 4 grids.
2. Write an addition problem for the shaded grids.

   0.23 + 0.23 + 0.23 + 0.23

3. Find the sum of your addition problem.

   0.92

4. Write a multiplication problem for your shaded picture.

   4 × 0.23

5. Find the product of your multiplication problem.

   0.92

LESSON 3-5

Puzzles, Twisters & Teasers

Maria's Twice Cut Cake

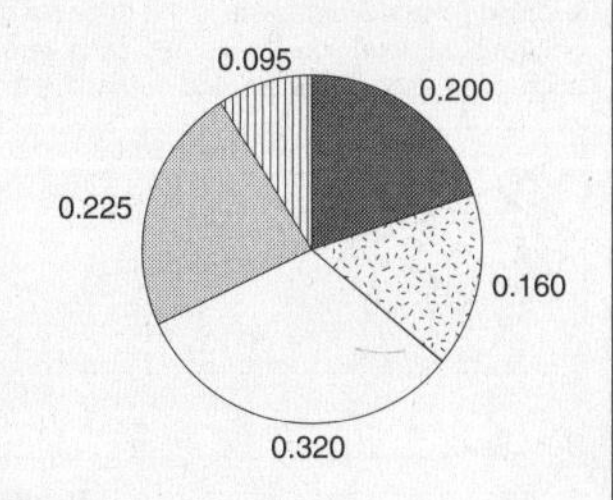

Today is Maria's birthday. Maria's coworkers, Bob, Makiko, José, and Edna, bought her a cake. Unfortunately, Maria and her coworkers are dieting, so none of them wanted a whole piece of cake. They each wanted a fraction of a piece. Maria wanted 0.2 of a piece, Bob wanted 0.16 of a piece, Makiko wanted 0.32 of a piece, José wanted 0.225 of a piece and Edna wanted just 0.095 of a piece. Because Maria was not a mathematician, she misunderstood her coworkers—she thought that they wanted the entire cake divided into 5 pieces. So she cut the cake as shown.

When the cake was cut, the coworkers received the wrong pieces. Maria got Bob's piece, Bob got Makiko's piece, Makiko got José's piece, José got Edna's piece, and Edna got Maria's piece.

Once they received their pieces, each coworker assumed they received a full piece of cake so each of them cut off the amount they had originally requested.

For example, Maria received Bob's 0.16 of the cake—so she ate her 0.2 piece of that. Mathematically she ate 0.2 x 0.16 = 0.032 piece of the cake. The question is: Who ate the most cake? List how much of the entire cake each of the five coworkers ate, then decide who ate the most cake.

Maria has 0.032 of a piece.

Bob has 0.0512

Makiko has 0.072

José has 0.021375

Edna has 0.019

So, who ate the most cake? Makiko

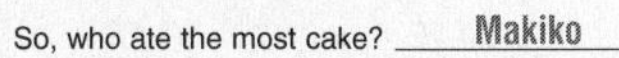

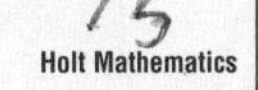

LESSON 3-6

Practice A

Dividing Decimals by Whole Numbers

Find each quotient.

1. 2.8 ÷ 4 — 0.7
2. 1.8 ÷ 2 — 0.9
3. 3.6 ÷ 6 — 0.6
4. 7.2 ÷ 9 — 0.8
5. 0.15 ÷ 3 — 0.05
6. 4.8 ÷ 8 — 0.6
7. 0.8 ÷ 4 — 0.2
8. 2.1 ÷ 7 — 0.3
9. 0.32 ÷ 4 — 0.08
10. 5.4 ÷ 9 — 0.6
11. 3.5 ÷ 5 — 0.7
12. 0.2 ÷ 2 — 0.1

Evaluate 2.4 ÷ *x* for each given value of *x*.

13. $x = 8$ — 0.3
14. $x = 2$ — 1.2
15. $x = 3$ — 0.8
16. $x = 4$ — 0.6
17. $x = 6$ — 0.4
18. $x = 12$ — 0.2

19. A six-pack of orange soda costs $4.20. How much does each can in the pack cost?

$0.70

20. It rained 2.7 inches in July and 2.1 inches in August. What was the average rainfall for those two months?

2.4 inches

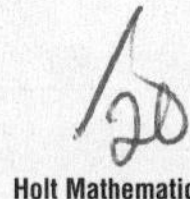

LESSON 3-6

Practice B

Dividing Decimals by Whole Numbers

Find each quotient.

1. 0.81 ÷ 9 — 0.09
2. 1.84 ÷ 4 — 0.46
3. 7.2 ÷ 6 — 1.2
4. 13.6 ÷ 8 — 1.7
5. 4.55 ÷ 5 — 0.91
6. 29.6 ÷ 8 — 3.7
7. 15.57 ÷ 9 — 1.73
8. 0.144 ÷ 12 — 0.012
9. 97.5 ÷ 3 — 32.5
10. 0.0025 ÷ 5 — 0.0005
11. 2.84 ÷ 8 — 0.355
12. 18.9 ÷ 3 — 6.3

Evaluate 2.094 ÷ *x* for each given value of *x*.

13. $x = 2$ — 1.047
14. $x = 4$ — 0.5235
15. $x = 12$ — 0.1745
16. $x = 20$ — 0.1047
17. $x = 15$ — 0.1396
18. $x = 30$ — 0.0698

19. There are three grizzly bears in the city zoo. Yogi weighs 400.5 pounds, Winnie weighs 560.35 pounds, and Nyla weighs 618.29 pounds. What is the average weight of the three bears?

526.38 pounds

20. The bill for dinner came to $75.48. The four friends decided to leave a $15.00 tip. If they shared the bill equally, how much will they each pay?

$22.62

LESSON 3-6

Practice C

Dividing Decimals by Whole Numbers

Find each quotient.

1. 2.36 ÷ 8 — 0.295
2. 0.1488 ÷ 3 — 0.0496
3. 72.654 ÷ 6 — 12.109
4. 8.523 ÷ 9 — 0.947
5. 115.8 ÷ 12 — 9.65
6. 0.952 ÷ 17 — 0.056
7. 46.545 ÷ 29 — 1.605
8. 14.795 ÷ 55 — 0.269
9. 0.2808 ÷ 75 — 0.003744

Evaluate *x* ÷ 6 for each value of *x*.

10. $x = 4.8$ — 0.8
11. $x = 0.54$ — 0.09
12. $x = 0.024$ — 0.004
13. $x = 1.08$ — 0.18
14. $x = 0.42$ — 0.07
15. $x = 0.0012$ — 0.0002

Evaluate.

16. $n \div 19$ for $n = 28.5$ — 1.5
17. $(6^2 + 1.35) \div c$ for $c = 5$ — 7.47
18. $4^3 - (0.81 \div x)$ for $x = 9$ — 63.91
19. $3.5t \div 4$ for $t = 19.36$ — 16.94

20. As of 2000, there were 281.42 million people in the United States. If the same number of people lived in each of the 50 states, what would have been the population of each state in 2000?

5.6284 million

21. In a gymnastics competition, Kim scored 9.4, 9.7, 9.9, and 9.8. Tamara scored 9.5, 9.2, 9.7, and 9.6. Who had the highest average score?

Kim

LESSON **3-6**
Reteach
Dividing Decimals by Whole Numbers

You can use decimal grids to help you divide decimals by whole numbers.

To divide 0.35 by 7, first shade in a decimal grid to show thirty-five hundredths.

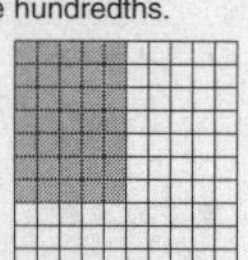

0.35 ÷ 7 means "divide 0.35 into 7 equal groups." Show this on the decimal grid.

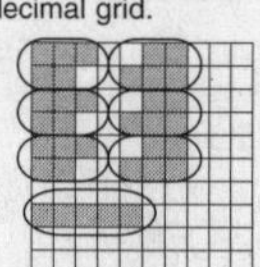

The number of units in each group is the quotient.

So, 0.35 ÷ 7 = 0.05.

/12

Use decimal grids to find each quotient.

1. 0.24 ÷ 4

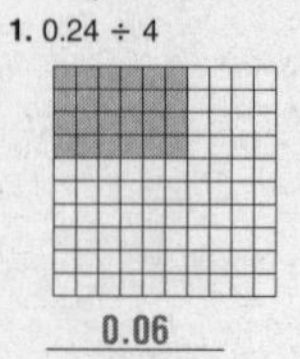

0.06

2. 0.48 ÷ 12

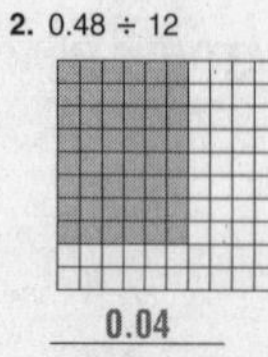

0.04

3. 0.50 ÷ 10

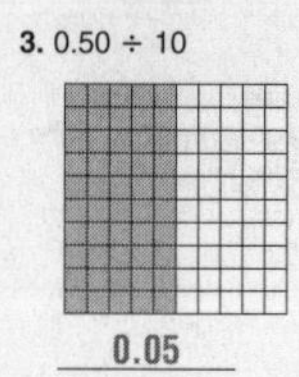

0.05

4. 0.98 ÷ 7

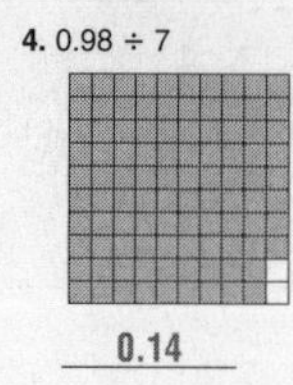

0.14

5. 0.6 ÷ 5 — 0.12
6. 0.78 ÷ 6 — 0.13
7. 0.99 ÷ 11 — 0.09
8. 0.32 ÷ 4 — 0.08

/8

LESSON **3-6**
Challenge
Get the Best Deal

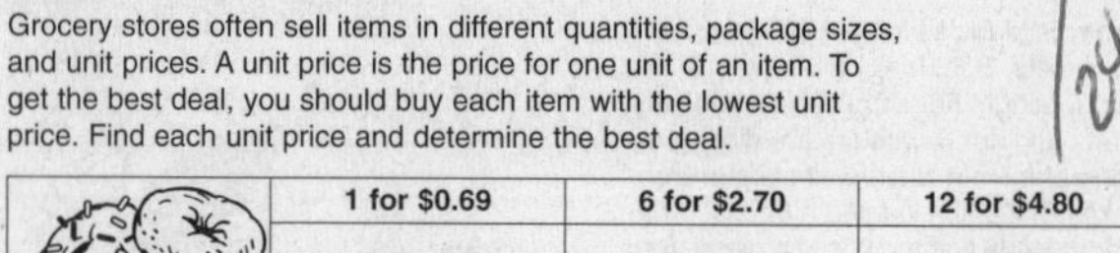

Grocery stores often sell items in different quantities, package sizes, and unit prices. A unit price is the price for one unit of an item. To get the best deal, you should buy each item with the lowest unit price. Find each unit price and determine the best deal.

	1 for $0.69	6 for $2.70	12 for $4.80
	Unit price (per pound) $0.69	Unit price (per pound) $0.45	Unit price (per pound) $0.40

Best deal: 12 for $4.80

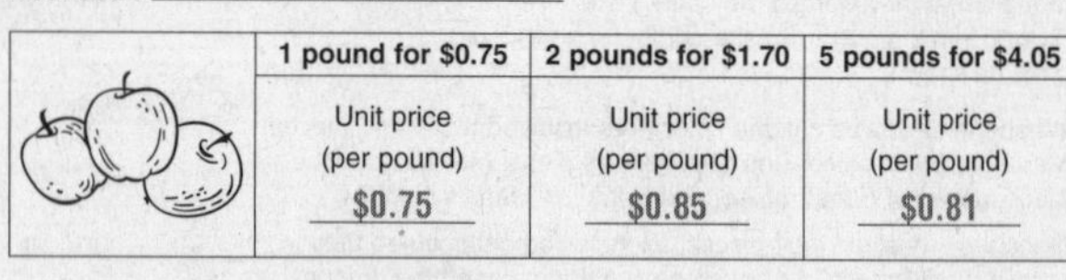

	1 pound for $0.75	2 pounds for $1.70	5 pounds for $4.05
	Unit price (per pound) $0.75	Unit price (per pound) $0.85	Unit price (per pound) $0.81

Best deal: 1 pound for $0.75

	6-ounce box for $1.98	12-ounce box for $3.72	16-ounce box for $5.28
	Unit price (per ounce) $0.33	Unit price (per ounce) $0.31	Unit price (per ounce) $0.33

Best deal: 12-ounce box for $3.72

	6-pack for $1.08	12-pack for $2.64	24-pack for $4.08
	Unit price (per can) $0.18	Unit price (per can) $0.22	Unit price (per can) $0.17

Best deal: 24-pack for $4.08

/16

LESSON **3-6**
Problem Solving
Dividing Decimals by Whole Numbers

Write the correct answer.

1. Four friends had lunch together. The total bill for lunch came to $33.40, including tip. If they shared the bill equally, how much did they each pay?

   $8.35

2. There are 7.2 milligrams of iron in a dozen eggs. Because there are 12 eggs in a dozen, how many milligrams of iron are in 1 egg?

   0.6 milligrams

3. Kyle bought a sheet of lumber 8.7 feet long to build fence rails. He cut the strip into 3 equal pieces. How long is each piece?

   2.9 feet

4. An albatross has a wingspan greater than the length of a car—3.7 meters! Wingspan is the length from the tip of one wing to the tip of the other wing. What is the length of each albatross wing (assuming wing goes from center of body)?

   1.85 meters

Circle the letter of the correct answer.

5. The City Zoo feeds its three giant pandas 181.5 pounds of bamboo shoots every day. Each panda is fed the same amount of bamboo. How many pounds of bamboo does each panda eat every day?
   - A 6.05 pounds
   - (B) 60.5 pounds
   - C 61.5 pounds
   - D 605 pounds

6. Emma bought 22.5 yards of cloth to make curtains for two windows in her apartment. She used the same amount of cloth on each window. How much cloth did she use to make each set of curtains?
   - F 1.125 yards
   - G 10.25 yards
   - (H) 11.25 yards
   - J 11.52 yards

7. Aerobics classes cost $153.86 for 14 sessions. What is the fee for one session?
   - (A) $10.99
   - B $1.99
   - C about $25.00
   - D about $20.00

8. An entire apple pie has 36.8 grams of saturated fat. If the pie is cut into 8 slices, how many grams of saturated fat are in each slice?
   - F 4.1 grams
   - G 0.46 grams
   - (H) 4.6 grams
   - J 4.11 grams

LESSON **3-6**
Reading Strategies
Use a Visual Tool

You can use a hundred grid to show division with decimals.

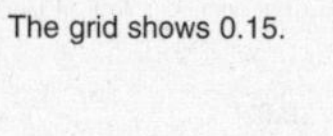

The grid shows 0.15. →

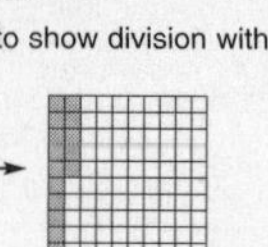

0.15 ÷ 3 means "separate 0.15 into 3 equal groups." →

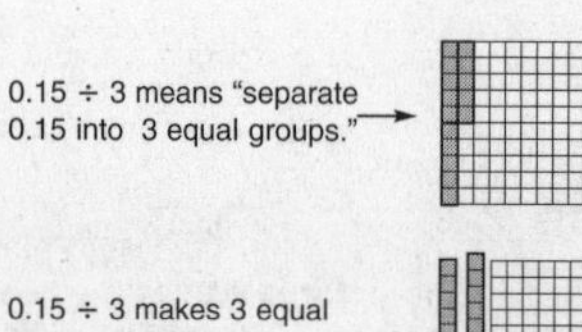

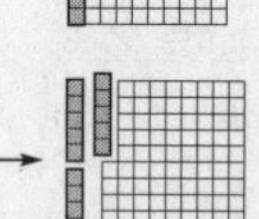

0.15 ÷ 3 makes 3 equal groups of 0.05. →

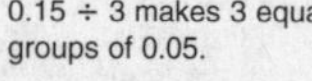

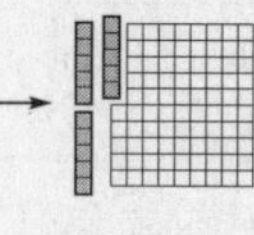

0.15 ÷ 3 = 0.05

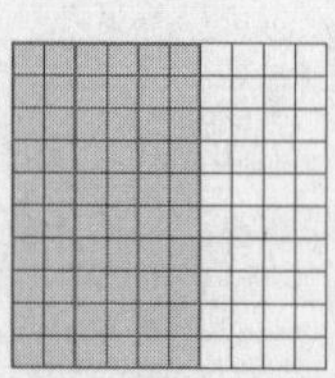

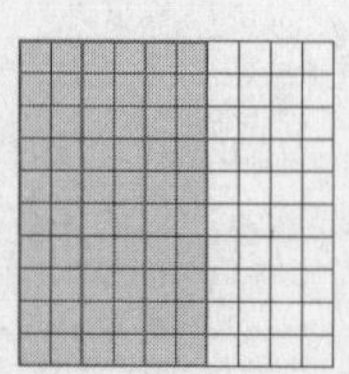

Use the grid to complete the Exercises 1–4.

1. Shade 0.60 of the grid.
2. Divide the grid into 3 equal groups.
3. Write the decimal amount in each of the 3 groups. 0.20
4. Write a division problem for the picture you have created.

   0.60 ÷ 3 = 0.20

14

LESSON 3-6

Puzzles, Twisters & Teasers

One Nation, Indivisible. . .

What state is round on both ends and raised in the middle?

We know the United States consists of fifty individual states. Often states want to make improvements to roads, state parks, or schools. They look to the federal government for monies to pay for those improvements.

For each of the improvements listed below, the government has decided which states will receive money. Divide the money for each item by the number of states receiving the money to find how much money each state gets.

ITEM	MONEY	RECIPIENT STATES	MONEY PER STATE
State Parks	\$193.45 per acre	CT, CO, WY, UT, AL	\$38.69
Farming	\$639.24 per acre	NE, KS, MO	\$213.08
Roads	\$534.36 per mile	WV, GA, MS, NC, OR, WA	\$89.06
Security	\$94.56 per government building	MD, PA	\$47.28
Schools	\$143.04 per classroom	LA, OK, ME, RI, KY, AL, WI, HI	\$17.88
Science/ Research	\$193.20 per laboratory	FL, TX, MA, CA, AZ	\$38.64
Museums	\$884.79 per display	NY, NM, NH	\$294.93

To answer the riddle, find the column labeled with largest amount of money and the row labeled with the smallest amount of money.

	\$305.17	\$296.40	\$294.93	\$213.08
\$22.04	Maryland	North Dakota	Oklahoma	Oregon
\$14.45	Ohio	Florida	Texas	Vermont
\$17.88	Iowa	Minnesota	New York	Louisiana
\$38.69	Pennsylvania	Arkansas	Maine	Arizona

So, what state do you think is round on both ends and raised in the middle?

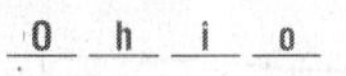

O h i o

LESSON 3-7

Practice A

Dividing by Decimals

Find each quotient.

1. 2.4 ÷ 0.4 — 6
2. 1.4 ÷ 0.2 — 7
3. 4.8 ÷ 0.6 — 8
4. 8.1 ÷ 0.9 — 9
5. 1.8 ÷ 0.3 — 6
6. 6.4 ÷ 0.8 — 8
7. 3.3 ÷ 0.3 — 11
8. 2.6 ÷ 1.3 — 2
9. 7.2 ÷ 1.2 — 6
10. 7.5 ÷ 1.5 — 5
11. 6.0 ÷ 0.5 — 12
12. 9.9 ÷ 1.1 — 9

Evaluate 4.8 ÷ x for each value of x.

13. x = 0.2 — 24
14. x = 0.4 — 12
15. x = 0.3 — 16
16. x = 0.6 — 8
17. x = 0.8 — 6
18. x = 1.2 — 4

19. Antonio spent \$5.60 on cashews. They cost \$1.40 per pound. How many pounds of cashews did Antonio buy?

    4 pounds

20. Over several months, a scientist measured a total of 6.3 inches of snow. The average snowfall each month was 2.1 inches. How many months did the scientist measure the snow?

    3 months

LESSON 3-7

Practice B

Dividing by Decimals

Find each quotient.

1. 9.0 ÷ 0.9 — 10
2. 29.6 ÷ 3.7 — 8
3. 10.81 ÷ 2.3 — 4.7
4. 10.5 ÷ 1.5 — 7
5. 15.36 ÷ 4.8 — 3.2
6. 9.75 ÷ 1.3 — 7.5
7. 20.4 ÷ 5.1 — 4
8. 37.5 ÷ 2.5 — 15
9. 9.24 ÷ 1.1 — 8.4
10. 16.56 ÷ 6.9 — 2.4
11. 28.9 ÷ 8.5 — 3.4
12. 14.35 ÷ 0.7 — 20.5

Evaluate x ÷ 1.2 for each value of x.

13. x = 40.8 — 34
14. x = 1.8 — 1.5
15. x = 10.8 — 9
16. x = 14.4 — 12
17. x = 4.32 — 3.6
18. x = 0.06 — 0.05

19. Anna is saving \$6.35 a week to buy a computer game that costs \$57.15. How many weeks will she have to save to buy the game?

    9 weeks

20. Ben ran a 19.5-mile race last Saturday. His average speed during the race was 7.8 miles per hour. How long did it take Ben to finish the race?

    2.5 hours

LESSON 3-7

Practice C

Dividing by Decimals

Find each quotient.

1. 4.75 ÷ 2.5 — 1.9
2. 34.04 ÷ 4.6 — 7.4
3. 10.0 ÷ 1.25 — 8
4. 283.62 ÷ 8.7 — 32.6
5. 168.75 ÷ 6.75 — 25
6. 0.1092 ÷ 0.013 — 8.4
7. 7.7293 ÷ 3.7 — 2.089
8. 97.206 ÷ 5.1 — 19.06
9. 0.489807 ÷ 0.081 — 6.047

Evaluate 15.65 ÷ x for each value of x.

10. x = 0.2 — 78.25
11. x = 0.4 — 39.125
12. x = 0.5 — 31.3
13. x = 0.8 — 19.5625
14. x = 1.6 — 9.78125
15. x = 2.5 — 6.26

Evaluate.

16. $n \div 7.8$ for $n = 26.988$ — 3.46
17. $(7^2 - 32.9) \div c$ for $c = 3.5$ — 4.6
18. $18.67 - (0.216 \div x)$ for $x = 0.02$ — 7.87
19. $4.4t \div 1.6$ for $t = 16.92$ — 46.53

20. The sum of two decimal numbers is 3.9. Their difference is 0.9, and their product is 3.6. What are the two numbers?

    2.4 and 1.5

21. The sum of two decimal numbers is 5.3. Their difference is 1.7, and their product is 6.3. What are the two numbers?

    3.5 and 1.8

LESSON **3-7**

Reteach

Dividing by Decimals

You can use powers of ten to help you divide a decimal by a decimal.

To divide 0.048 by 0.12, first multiply each number by the least power of ten that makes the divisor a whole number.

$0.048 \div 0.12$

$0.12 \cdot 10^2 = 12$ Move the decimal point 2 places to the right.

$0.048 \cdot 10^2 = 4.8$ Move the decimal point 2 places to the right.

Then divide.

$4.8 \div 12$ **Step 1:** Divide as you would divide a whole number by a whole number.

$$\begin{array}{r} 0.4 \\ 12\overline{)4.8} \\ \underline{4\,8} \\ 0 \end{array}$$

Step 2: Think $48 \div 12 = 4$.

Step 3: Bring the decimal into the quotient and add a zero placeholder if necessary.

So, $0.048 \div 0.12 = 0.4$.

Find each quotient.

1. $0.7\overline{)0.42}$ **0.6**
2. $0.08\overline{)0.4}$ **5**
3. $0.5\overline{)0.125}$ **0.25**
4. $0.02\overline{)0.3}$ **15**
5. $0.4\overline{)0.08}$ **0.2**
6. $0.9\overline{)0.63}$ **0.7**
7. $0.008\overline{)0.4}$ **50**
8. $0.04\overline{)0.032}$ **0.8**
9. $0.3\overline{)0.06}$ **0.2**
10. $0.04\overline{)0.2}$ **5**
11. $0.007\overline{)4.9}$ **700**
12. $0.6\overline{)0.012}$ **0.02**

LESSON **3-7**

Challenge

Cutting Decimals

The strips of cloth below need to be cut into equal pieces of given lengths. Draw lines on each strip of cloth to show how many pieces will be cut.

1. Total Length: 9.8 yards Piece Length: 1.4 yards

Students should draw 6 lines across the cloth to cut it into 7 equal pieces.

2. Total Length: 2.5 yards Piece Length: 0.5 yards

Students should draw 4 lines across the cloth to cut it into 5 equal pieces.

3. Total Length: 10.2 yards Piece Length: 1.7 yards

Students should draw 5 lines across the cloth to cut it into 6 equal pieces.

4. Total Length: 6.4 yards Piece Length: 0.8 yards

Students should draw 7 lines across the cloth to cut it into 8 equal pieces.

5. Total Length: 13.6 yards Piece Length: 3.4 yards

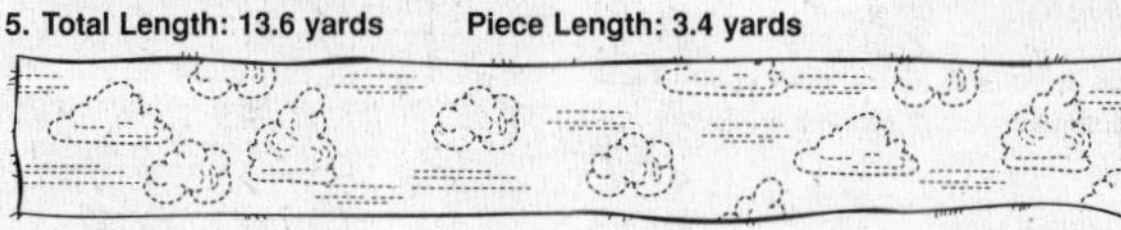

Students should draw 3 lines across the cloth to cut it into 4 equal pieces.

LESSON **3-7**

Problem Solving

Dividing by Decimals

Write the correct answer.

1. Jamal spent $6.75 on wire to build a rabbit hutch. Wire costs $0.45 per foot. How many feet of wire did Jamal buy?

   15 feet

2. Peter drove 195.3 miles in 3.5 hours. On average, how many miles per hour did he drive?

   55.8 miles per hour

3. Lisa's family drove 830.76 miles to visit her grandparents. Lisa calculated that they used 30.1 gallons of gas. How many miles per gallon did the car average?

   27.6 miles per gallon

4. A chef bought 84.5 pounds of ground beef. He uses 0.5 pound of ground beef for each hamburger. How many hamburgers can he make?

   169 hamburgers

Circle the letter of the correct answer.

5. Mark earned $276.36 for working 23.5 hours last week. He earned the same amount of money for each hour that he worked. What is Mark's hourly rate of pay?
   - A $1.17
   - B $10.76
   - (C) $11.76
   - D $117.60

6. Alicia wants to cover a section of her wall that is 2 feet wide and 12 feet long with mirrors. Each mirror tile is 2 feet wide and 1.5 feet long. How many mirror tiles does she need to cover that section?
   - F 4 tiles
   - G 6 tiles
   - (H) 8 tiles
   - J 12 tiles

7. John ran the city marathon in 196.5 minutes. The marathon is 26.2 miles long. On average, how many miles per hour did John run the race?
   - A 7 miles per hour
   - B 6.2 miles per hour
   - (C) 8 miles per hour
   - D 8.5 miles per hour

8. Shaneeka is saving $5.75 of her allowance each week to buy a new camera that costs $51.75. How many weeks will she have to save to have enough money to buy it?
   - (F) 9 weeks
   - G 9.5 weeks
   - H 8.1 weeks
   - J 8 weeks

LESSON **3-7**

Reading Strategies

Make Predictions

Study the examples below. Look for patterns in the divisor and quotient.

Dividend		Divisor		Quotient
400	÷	20	=	20
400	÷	2	=	200
400	÷	0.2	=	2,000
400	÷	0.02	=	20,000

As the divisor is divided by 10, the quotient is multiplied by 10.

Use the information above to answer the Exercises 1–3.

1. Predict the divisor for the next problem in this pattern.

   0.002

2. Predict the quotient for the next problem in this pattern.

   200,000

3. Write the next division problem and quotient for this pattern.

   400 ÷ 0.002 = 200,000

Study the pattern created by these division problems. Use the pattern to answer Exercises 4–6.

Dividend		Divisor		Quotient
900	÷	30	=	30
900	÷	3	=	300
900	÷	0.3	=	3,000

4. Predict the next divisor in this pattern.

   0.03

5. Predict the next quotient in this pattern.

   30,000

6. Write the division problem and quotient that you predict would come next.

   900 ÷ 0.03 = 30,000

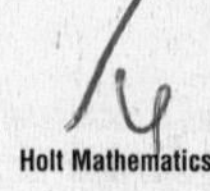

LESSON 3-7 Puzzles, Twisters & Teasers
Bargain Hunter

You are a professional shopper. You've been asked to use your shopping skills to find the best buy among the following cereals. You must determine which of these cereals costs the least per ounce.

Type of Cereal	Size of Box	Price per Box
Frostee O's	22.5 oz	$12.60
Super Sugar Loops	9.9 oz	$2.97
Marshmallow Bonanza	22.8 oz	$3.42
Tyrannosaurus Rings	6.6 oz	$3.30
Wheat and Rice Explosion	35.6 oz	$3.56
Giant Frosted Oatees	10.0 oz	$6.10
Ned's Enormous O's of Sugar	15.5 oz	$1.86

Type of Cereal	Price per Oz
Frostee O's	$0.56
Super Sugar Loops	$0.30
Marshmallow Bonanza	$0.15
Tyrannosaurus Rings	$0.50
Wheat and Rice Explosion	$0.10
Giant Frosted Oatees	$0.61
Ned's Enormous O's of Sugar	$0.12

Do you know what lies at the bottom of the ocean and twitches?

To find out, fill in the capital letters from the *second* best bargain cereal in order in the spaces below. Then fill in the capital letters in order from the *best* bargain cereal and you should have the answer!

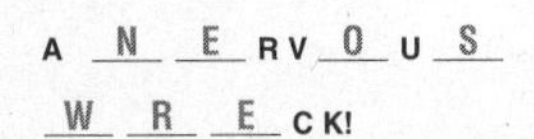

A N E RV O U S

W R E CK!

14

LESSON 3-8 Practice A
Interpret the Quotient

Circle the letter of the correct answer.

1. Hamburger rolls come in packs of 8. How many packs should you buy to have 60 rolls?
   (A) 8
   B 6
   C 5
   D 7

2. Each pack of hamburger rolls costs $1.50. How many packs can you buy with $8.00?
   F 6
   (G) 5
   H 4
   J 8

3. How many 0.6-pound hamburgers can you make with 7.8 pounds of ground beef?
   (A) 13
   B 14
   C 10
   D 16

4. You spend a total of $5.10 for 3 pounds of ground beef. How much does the ground beef cost per pound?
   F $0.70
   G $0.17
   H $15.30
   (J) $1.70

Write the correct answer.

5. Four friends equally shared the cost of buying supplies for the class picnic. The supplies cost a total of $12.40. How much did they each pay?
   $3.10

6. In all, 20 people are going to the picnic. Each van seats 6 people. How many vans are needed to take everyone to the picnic?
   4 vans

7. Plastic forks come in packs of 6. If you need 40 forks for the picnic, how many packs should you buy?
   7 packs

8. You spent a total of $9.60 on paper plates for the picnic. Each pack costs $1.20. How many packs of paper plates did you buy?
   8 packs

LESSON 3-8 Practice B
Interpret the Quotient

Circle the letter of the correct answer.

1. You spent a total of $6.75 for 15 yards of ribbon. How much did the ribbon cost per yard?
   A $0.50
   (B) $0.45
   C $1.35
   D $1.45

2. Buttons come in packs of 12. How many packs should you buy if you need 100 buttons?
   F 10
   G 8
   (H) 9
   J 12

3. Your sewing cabinet has compartments that hold 8 spools of thread each. You have 50 spools of thread. How many compartments can you fill?
   (A) 6
   B 7
   C 5
   D 8

4. You spent a total of $35.75 for velvet cloth. Each yard of the velvet costs $3.25. How many yards did you buy?
   F 10
   G 10.5
   (H) 11
   J 11.5

Write the correct answer.

5. You used a total of 67.5 yards of cotton material to make costumes for the play. Each costume used 11.25 yards of cloth. How many costumes did you make?
   6 costumes

6. You are saving $17.00 each week to buy a new sewing machine that costs $175.50. How many weeks will you have to save to have enough money to buy the sewing machine?
   11 weeks

7. Sequins come in packs of 75. You use 12 sequins on each costume. If you have one pack of sequins, how many costumes can you make?
   6 costumes

8. You pay $26.28 for a subscription to *Sewing Magazine*. You get an issue every month for a year. How much does each issue cost?
   $2.19

LESSON 3-8 Practice C
Interpret the Quotient

Write the correct answer.

1. You live in Detroit, Michigan. You and your parents will be driving to Cincinnati, Ohio, to visit your grandparents. The trip is a total of 264.1 miles. Your family car averages 27.8 miles per gallon. How many gallons of gas will you use on the trip?
   9.5 gallons

2. You plan on taking 100 photographs during your trip. You want to choose one kind of film to buy, either a 24-photo roll, or an 18-photo roll. How many packs of each would you need to buy?
   5 rolls of the 24-photo or 6 rolls of the 18-photo film

3. The gas tank of your parents' car holds 12.9 gallons of gas. Because their car averages 27.8 miles per gallon, how many times will they have to fill up the gas tank to drive to Cincinnati and back?
   2 times

4. The first time you stop at a gas station during your trip, your parents spend $13.80 for 11.5 gallons of gas. How much does the gas cost per gallon?
   $1.20 per gallon

5. You stop for lunch during the trip and spend a total of $11.38. Your cheeseburger costs $3.25. Both of your parents have a slice of pizza that each cost $2.19. You all get the same drink. How much did each drink cost?
   $1.25

6. During the trip your parents drive 211.05 miles on different highways. You calculate that you spent 3.5 hours driving on highways. What was the average speed your parents drove during that part of the trip?
   60.3 miles per hour

7. You saved for 16 weeks before your trip to buy a present for your grandparents that cost $67.36. How much did you save each week?
   $4.21

8. Your parents budgeted $20 a month for long-distance calls. It costs $0.37 per minute for long distance calls. How many minutes can you talk to your grandparents each month?
   54 minutes

LESSON 3-8 **Reteach**

Interpret the Quotient

There are three ways the decimal part of a quotient can be interpreted when you solve a problem.

If the question asks for an exact number, use the entire quotient.
If the question asks how many whole groups are needed to put the dividend into a group, round the quotient up to the next whole number.
If the question asks how many whole groups can be made when you divide, drop the decimal part of the quotient.

To interpret the quotient, decide what the question is asking.

In the school library, there are tables that seat 4 students each. If there are 30 students in a class, how many tables are needed to seat all of the students?

To solve, divide 30 by 4.

$30 \div 4 = 7.5$

The question is asking how many tables (whole groups) are needed to put all of the students in the class (dividend) into a group.

So, round 7.5 up to the next whole number.

8 tables are needed to seat all of the students.

Interpret the quotient to solve each problem.

1. A recipe that serves 6 requires 9 cups of milk. How much milk is needed for each serving?

   1.5 cups are needed for each serving.

2. A storage case holds 24 model cars. Marla has 84 model cars. How many storage cases does she need to store all of her cars?

   Marla needs 4 cases.

3. Kenny has $4.25 to spend at the school carnival. If game tickets are $0.50 each, how many games can Kenny play?

   Kenny can play 8 games.

LESSON 3-8 **Challenge**

Plan a Party!

You are in charge of buying supplies for the class party. There are 30 students in your class. Use the party supply store advertisement below to plan what to buy. After you pay for all the items, the total cost will be divided evenly among all the students.

Shopping List

Item	Number of Items You Want Per Person	Number of Packs to Buy	Number of Left Over Items	Total Price of Items
Invitations	1	4	2	$7.40
Paper plates	1	2	20	$5.60
Plastic cups	2	6	0	$11.70
Paper napkins	2	1	15	$3.65
Plastic forks	1	3	6	$6.15
			Grand Total Price:	$34.50
			Cost Per Student:	$1.15

LESSON 3-8 **Problem Solving**

Interpret the Quotient

Write the correct answer.

1. Five friends split a pizza that costs $16.75. If they shared the bill equally, how much did they each pay?

   $3.35

2. There are 45 choir members going to the recital. Each van can carry 8 people. How many vans are needed?

   6 vans

3. Tara bought 150 beads. She needs 27 beads to make each necklace. How many necklaces can she make?

   5 necklaces

4. Cat food costs $2.85 for five cans. Ben only wants to buy one can. How much will it cost?

   $0.57

Circle the letter of the correct answer.

5. Tennis balls come in cans of 3. The coach needs 50 tennis balls for practice. How many cans should he order?
   - A 16 cans
   - (B) 17 cans
   - C 18 cans
   - D 20 cans

6. The rainfall for three months was 4.6 inches, 3.5 inches, and 4.2 inches. What was the average monthly rainfall during that time?
   - F 41 inches
   - G 12.3 inches
   - H 4.3 inches
   - (J) 4.1 inches

7. Tom has $15.86 to buy marbles that cost $1.25 each. He wants to know how many marbles he can buy. What should he do after he divides?
   - (A) Drop the decimal part of the quotient when he divides.
   - B Drop the decimal part of the dividend when he divides.
   - C Round the quotient up to the next highest whole number to divide.
   - D Use the entire quotient of his division as the answer.

8. Mei needs 135 hot dog rolls for the class picnic. The rolls come in packs of 10. She wants to know how many packs to buy. What should she do after she divides?
   - F Drop the decimal part of the quotient when she divides.
   - G Drop the decimal part of the dividend when she divides.
   - (H) Round the quotient up to the next highest whole number.
   - J Use the entire quotient of her division as the answer.

LESSON 3-8 **Reading Strategies**

Use Context

How the decimal portion of the quotient in a division problem is used depends upon the situation.

Situation 1 74 students are going on a field trip in cars. Each car can carry 5 students. How many cars are needed?

Divide 74 by 5. → $74 \div 5 = 14.8$ cars

Reasoning 14 cars will not be enough for all students. You need 15 cars. The quotient 14.8 needs to be **rounded up** to 15 in this situation.

Situation 2 How many 8 oz servings are in a 44 oz can of juice?

Divide 44 by 8. → $44 \div 8 = 5.5$ servings

Reasoning There are 5 full 8 oz servings in the can. The 0.5 serving is not 8 ounces. The quotient 5.5 is **rounded down** to 5 in this situation.

Situation 3 4 boys mowed a lawn for $35. How much money should each boy receive to share the money equally?

Divide $35 by 4. → $35 ÷ 4 = $8.75

Reasoning The exact quotient of $8.75 states what each boy should receive. The **exact quotient** of $8.75 makes sense.

Tell whether you would round the quotient up, round the quotient down, or leave the exact quotient for each. Write to explain your choice.

1. You need 8 inches of ribbon to make a bow. How many bows can you make with 50 inches of ribbon? $50 \div 8 = 6.25$

   round down; You can make only 6 bows with 50 inches of ribbon.

2. Each lunch table seats 10 children. There are 155 children in the cafeteria for each lunch period. How many tables are needed? $155 \div 10 = 15.5$

   round up; You need another lunch table to seat everyone.

LESSON **3-8**

Puzzles, Twisters & Teasers

To Dine or Not to Dine?

John loves to have his friends over for dinner. He is concerned that he will not have enough food.

He knows that he has 8 cups of soup, 14 pounds of lasagna, 2.85 pounds of salad, and 7.3 gallons of ice cream. John knows that, on average, each guest eats about 0.75 cups of soup, 1.2 pounds of lasagna, 0.3 pounds of salad, and 0.6 pint of ice cream. Use these numbers to help John decide how many people he can invite for dinner.

Soup **10.66**

Salad **9.5**

Lasagna **11.66**

Ice Cream **12.16**

7	T
8	O
9	U
10	M
11	L
12	N

Number of people John should invite: **9**

Do you know what John's gossiping friend had during dinner?

Find the letter that is paired with the number of people that John should invite and add it to the letter bank. Also, add one to the number of people John can invite, find the matching letter in the table above and add that letter to the letter bank as well.

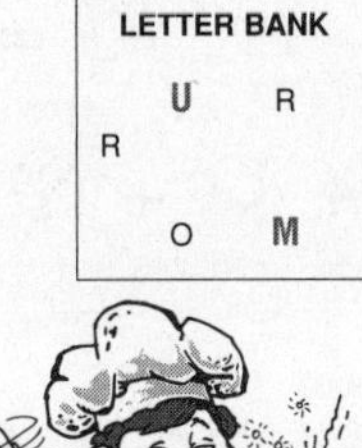

LETTER BANK

U R

R

O M

Rearrange the letters in the letter bank to find the answer to the riddle.

He had a great sense of

R U M O R.

LESSON **3-9**

Practice A

Solving Decimal Equations

Circle the letter of the correct answer.

1. If $7.2 + x = 9.7$, what is the value of x?
   A 16.9
   (B) 2.5
   C 2.9
   D 16.5

2. If $n \div 4 = 0.4$, what is the value of n?
   F 4.4
   G 0.1
   (H) 1.6
   J 10

Solve each equation. Check your answer.

3. $a - 0.4 = 1.3$
   $a = 1.7$; $1.7 - 0.4 = 1.3$

4. $2n = 1.8$
   $n = 0.9$; $2 \cdot 0.9 = 1.8$

5. $0.8 + x = 1.3$
   $x = 0.5$; $0.8 + 0.5 = 1.3$

6. $p \div 4 = 0.7$
   $p = 2.8$; $2.8 \div 4 = 0.7$

7. $w - 2.3 = 1.2$
   $w = 3.5$; $3.5 - 2.3 = 1.2$

8. $0.5q = 0.25$
   $q = 0.5$; $0.5 \cdot 0.5 = 0.25$

9. $5.7 + s = 6.0$
   $s = 0.3$; $5.7 + 0.3 = 6.0$

10. $b \div 3 = 0.6$
    $b = 1.8$; $1.8 \div 3 = 0.6$

11. $t - 3.1 = 1.6$
    $t = 4.7$; $4.7 - 3.1 = 1.6$

12. $3y = 1.5$
    $y = 0.5$; $3 \cdot 0.5 = 1.5$

13. The length of a photograph is 2.1 inches, and the width is 3 inches. Solve the equation $a \div 2.1 = 3$ to find the area of the photograph.
    $a = 6.3$ square inches

14. It costs \$0.90 to enlarge each photograph. If you want to enlarge 5 photos, how much will it cost in all?
    \$4.50

LESSON **3-9**

Practice B

Solving Decimal Equations

Solve each equation. Check your answer.

1. $a - 2.7 = 4.8$
   $a = 7.5$; $7.5 - 2.7 = 4.8$

2. $b \div 7 = 1.9$
   $b = 13.3$; $13.3 \div 7 = 1.9$

3. $w - 6.5 = 3.8$
   $w = 10.3$; $10.3 - 6.5 = 3.8$

4. $p \div 0.4 = 1.7$
   $p = 0.68$; $0.68 \div 0.4 = 1.7$

5. $4.5 + x = 8$
   $x = 3.5$; $4.5 + 3.5 = 8$

6. $b \div 3 = 2.5$
   $b = 7.5$; $7.5 \div 3 = 2.5$

7. $7.8 + s = 15.2$
   $s = 7.4$; $7.8 + 7.4 = 15.2$

8. $1.63q = 9.78$
   $q = 6$; $1.63 \cdot 6 = 9.78$

9. $0.05 + x = 2.06$
   $x = 2.01$; $0.05 + 2.01 = 2.06$

10. $1.7n = 2.38$
    $n = 1.4$; $1.7 \cdot 1.4 = 2.38$

11. $t - 6.08 = 12.59$
    $t = 18.67$; $18.67 - 6.08 = 12.59$

12. $9q = 16.2$
    $q = 1.8$; $9 \cdot 1.8 = 16.2$

13. $w - 8.9 = 10.3$
    $w = 19.2$; $19.2 - 8.9 = 10.3$

14. $1.4n = 3.22$
    $n = 2.3$; $1.4 \cdot 2.3 = 3.22$

15. $t - 12.7 = 0.8$
    $t = 13.5$; $13.5 - 12.7 = 0.8$

16. $3.8 + a = 6.5$
    $a = 2.7$; $3.8 + 2.7 = 6.5$

17. The distance around a square photograph is 12.8 centimeters. What is the length of each side of the photograph?
    3.2 centimeters

18. You buy two rolls of film for \$3.75 each. You pay with a \$10 bill. How much change should you get back?
    \$2.50

/18

LESSON **3-9**

Practice C

Solving Decimal Equations

Solve each equation. Check your answer.

1. $a - 0.089 = 12.5$
   $a = 12.589$; $12.589 - 0.089 = 12.5$

2. $b \div 2.8 = 4.7$
   $b = 13.16$; $13.16 \div 2.8 = 4.7$

3. $w - 12.76 = 15.8$
   $w = 28.56$; $28.56 - 12.76 = 15.8$

4. $p \div 0.07 = 4.76$
   $p = 0.3332$; $0.3332 \div 0.07 = 4.76$

5. $1.7806 + x = 2.009$
   $x = 0.2284$; $1.7806 + 0.2284 = 2.009$

6. $b \div 6.5 = 9.7$
   $b = 63.05$; $63.05 \div 6.5 = 9.7$

7. $8.3044 + s = 16.01$
   $s = 7.7056$; $8.3044 + 7.7056 = 16.01$

8. $0.009q = 0.765$
   $q = 85$; $0.009 \cdot 85 = 0.765$

9. $23.764 + x = 30.5$
   $x = 6.736$; $23.764 + 6.736 = 30.5$

10. $1.7n + 3.8 = 7.71$
    $n = 2.3$; $1.7 \cdot 2.3 + 3.8 = 7.71$

11. $t - (8^2 + 0.36) = 0.5$
    $t = 64.86$; $64.86 - 64.36 = 0.5$

12. $9q + 0.2q = 27.6$
    $q = 3$; $9.2 \cdot 3 = 27.6$

13. $8.4w - 0.67 = 15.29$
    $w = 1.9$; $(8.4 \cdot 1.9) - 0.67 = 15.29$

14. $1.4n + 4^3 = 85$
    $n = 15$; $(1.4 \cdot 15) + 64 = 85$

15. Andy bought 15 pounds of apples for \$0.69 a pound, and 7.2 pounds of grapes for \$3.65 a pound. He paid for the entire purchase with a \$50 bill. How much change did he get back? He wants to make 25 pounds of fruit salad for the party. How much more fruit does he need?
    \$13.37; 2.8 pounds

16. Brenda's garden is 6.5 feet long and 8.76 feet wide. She uses half of the garden for vegetables. Then she equally divides the rest of the garden between flowers and herbs. How large is her vegetable garden? How much land does she use for growing flowers?
    28.47 square feet;
    14.235 square feet

LESSON 3-9 **Reteach**

Solving Decimal Equations

You can write related equations for addition and subtraction equations.

$7.4 + 6.2 = 13.6$ $\quad$ $13.6 - 6.2 = 7.4$

Use related equations to solve each of the following.

A. $x + 4.5 = 7.9$
Think: $7.9 - 4.5 = x$
$x = 3.4$

Check $x + 4.5 = 7.9$
$3.4 + 4.5 \stackrel{?}{=} 7.9$ substitute
$7.9 = 7.9$

B. $x - 0.08 = 6.2$
Think: $6.2 + 0.08 = x$
$x = 6.28$

Check $x - 0.08 = 6.2$
$6.28 - 0.08 \stackrel{?}{=} 6.2$ substitute
$6.2 = 6.2$

Use related facts to solve each equation. Then check each answer.

1. $x + 8.7 = 12.9$ — $x = 4.2$
2. $x + 8.4 = 16.6$ — $x = 8.2$
3. $x - 2.65 = 7.8$ — $x = 10.45$
4. $x - 0.8 = 2.3$ — $x = 3.1$

You can write related equations for multiplication and division equations.

$3.2 \cdot 2.4 = 7.68$ $\quad$ $7.68 \div 2.4 = 3.2$

Use related equations to solve each of the following.

C. $3x = 1.5$
Think: $1.5 \div 3 = x$
$x = 0.5$

Check: $3x = 1.5$
$3 \cdot 0.5 \stackrel{?}{=} 1.5$ substitute
$1.5 = 1.5$

D. $x \div 6 = 1.2$
Think: $1.2 \cdot 6 = x$
$x = 7.2$

Check: $x \div 6 = 1.2$
$7.2 \div 6 \stackrel{?}{=} 1.2$ substitute
$1.2 = 1.2$

Use related facts to solve each equation. Then check each answer.

5. $x \div 3 = 6.3$ — $x = 18.9$
6. $x \div 0.2 = 3.4$ — $x = 0.68$
7. $7x = 4.2$ — $x = 0.6$
8. $5x = 4.5$ — $x = 0.9$

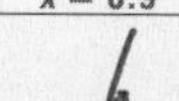

LESSON 3-9 **Challenge**

Playing Weight

In professional sports, each ball has a maximum, or greatest, weight allowed in play. The lightest official weight for a table tennis ball is only 0.009 ounces. Use the equations below to find the maximum weights, in ounces, of some other sports' balls.

[table tennis ball] + 1.601 = [golf ball]

[tennis ball] − [golf ball] = 0.45

[billiard ball] − 1.87 = [tennis ball] • 2

[volleyball] = [billiard ball] + 3.87

[golf ball] + [tennis ball] + [volleyball] + [billiard ball] + 236.03 = [bowling ball]

Golf Ball
Weight: 1.61 ounces

Tennis Ball
Weight: 2.06 ounces

Billiard Ball
Weight: 5.99 ounces

Volleyball
Weight: 9.86 ounces

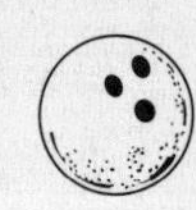

Bowling Ball
Weight: 255.55 ounces

LESSON 3-9 **Problem Solving**

Solving Decimal Equations

Write the correct answer.

1. Bee hummingbirds weigh only 0.0056 ounces. They have to eat half their body weight every day to survive. How much food does a bee hummingbird have to eat each day?
   0.0028 ounces
2. The desert locust, a type of grasshopper, can jump 10 times the length of its body. The locust is 1.956 inches long. How far can it jump in one leap?
   19.56 inches
3. In 1900, there were about 1.49 million people living in California. In 2000, the population was 33.872 million. How much did the population grow between 1900 and 2000?
   by 32.382 million
4. Juanita has $567.89 in her checking account. After she deposited her paycheck and paid her rent of $450.00, she had $513.82 left in the account. How much was her paycheck?
   $395.93

Circle the letter of the correct answer.

5. The average body temperature for people is 98.6°F. The average body temperature for most dogs is 3.4°F higher than for people. The average body temperature for cats is 0.5°F lower than for dogs. What is the normal body temperature for dogs and cats?
   A dogs: 101.5°F; cats 102°F
   (B) dogs: 102°F; cats 101.5°F
   C dogs: 102.5°F; cats 103°F
   D dogs: 102.5°F; cats 102.5°F
6. Seattle, Washington, is famous for its rainy climate. Winter is the rainiest season there. From November through December the city gets an average of 5.85 inches of rain each month. Seattle usually gets 6 inches of rain in December. What is the city's average rainfall in November?
   F 6 inches
   G 5.925 inches
   H 5.8 inches
   (J) 5.7 inches
7. The equation to convert from Celsius to Kelvin degrees is K = 273.16 + C. If it is 303.66°K outside, what is the temperature in Celsius degrees?
   A 576.82°C
   (B) 30.5°C
   C 305°C
   D 257.68°C
8. The distance around a square mirror is 6.8 feet. Which of the following equations finds the length of each side of the mirror?
   F $6.8 - x = 4$
   G $x \div 4 = 6.8$
   (H) $4x = 6.8$
   J $6.8 + 4 = x$

LESSON 3-9 **Reading Strategies**

Use a Flowchart

This flowchart can help you work with decimal equations.

1. How do I read this equation?
↓
2. What operation is shown in this equation?
↓
3. What is the inverse operation for this equation?

Adding or Subtracting Decimals

$x + 9.7 = 15.4$ ← **1.** Read "x plus 9.7 equals 15.4."
$x + \mathbf{9.7} = 15.4$ ← **2.** Addition is shown.
$x + 9.7 - \mathbf{9.7} = 15.4 - \mathbf{9.7}$ ← **3.** The inverse operation is subtraction.

Use this equation to complete Exercises 1–2: $n - 4.5 = 6.3$.

1. Write in words how you read the equation.
   n minus 4.5 equals 6.3.
2. What operation is shown in this equation? What is the inverse of that operation?
   subtraction; addition

Use this equation to complete Exercises 3–4: $w + 9 = 4.8$.

3. Write in words how you read this equation.
   w plus 9 equals 4.8.
4. What operation is shown in this equation? What is the inverse of that operation?
   addition; subtraction

LESSON 3-9

Puzzles, Twisters & Teasers

Super Cross

The Super Cross is the empty grid below. Fill in the grid with the letters from the equations below to answer the riddle.

First, solve all of the equations. Then enter the letter with the highest value in the top left corner. The next highest goes in the space to its right and then continue filling in the numbers from highest to lowest, left to right and top to bottom.

Read the answer to the riddle by reading down the rightmost column, then down the center column, then down the leftmost column.

$4.9b = 0.637$

$l - 54 = 45.1$

$\frac{h}{5} = 6.5$

$i + 9.3 = 15.6$

$4s = 2.8$

$65.7 - t = 24.9$

$a + 4.5 = 21.7$

$\frac{e}{5.6} = 3.2$

$34.2s = 3.42$

b	0.13
l	99.1
h	32.5
i	6.3
s	0.7
t	40.8
a	17.2
e	17.92
s	0.1

l	t	h
e	a	i
s	b	s

Do you know why the waiter was good at multiplication?

Because he knew

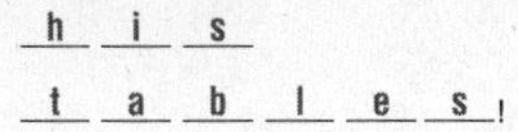

h i s

t a b l e s !

10